Robert T. Jones
215 Groveland Ave.
Summer 1927.

For Tom,

Happy Birthday to a
fine person — good
luck in your doing.

your friend,

Charlie

PLATE VIII

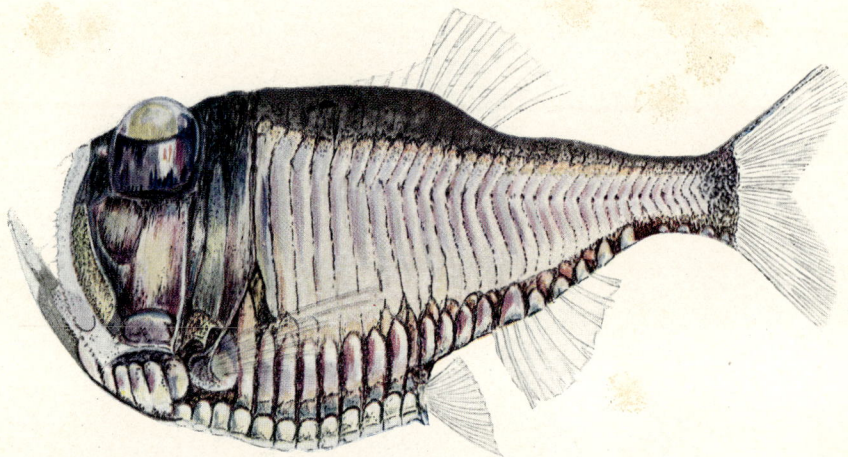

FIG. A

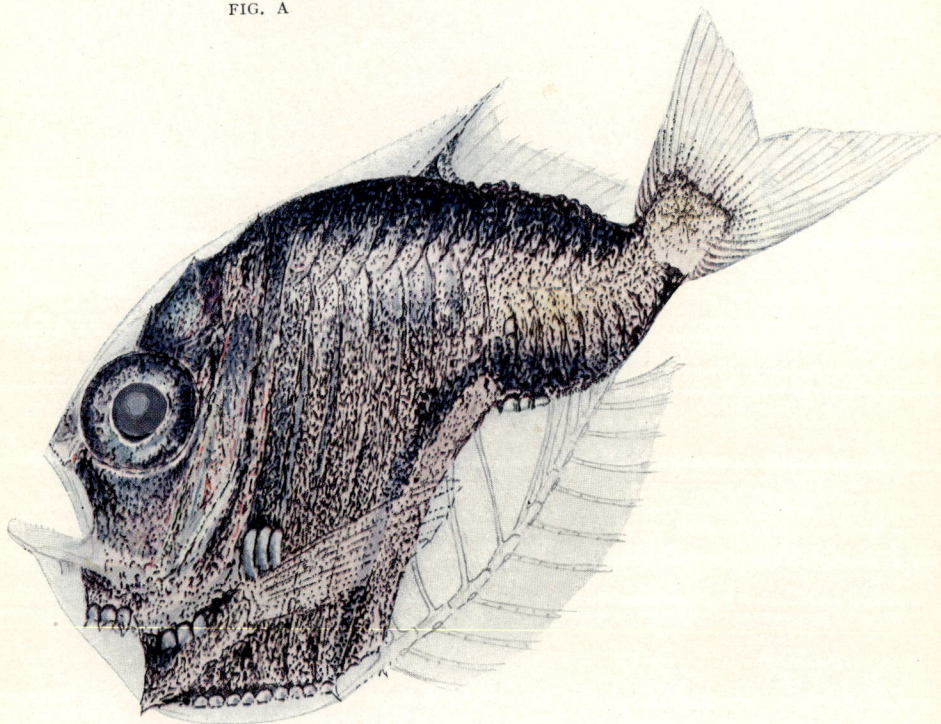

FIG. B

FIG. A. *Argyropelecus*. A LUMINESCENT DEEP-
SEA FISH WITH EYES DIRECTED UPWARD

(Twice natural size)

FIG. B. *Sternoptyx*. A COMMON DEEP-SEA FISH
WITH PART OF THE INNER FIN SKELETON
VISIBLE OUTSIDE OF THE BODY

(Twice natural size)

THE
ARCTURUS ADVENTURE

An Account of the New York Zoological Society's
First Oceanographic Expedition

BY

WILLIAM BEEBE

Director of the Department of Tropical Research

WITH 77 ILLUSTRATIONS FROM COLORED PLATES,
PHOTOGRAPHS AND MAPS

Published under the Auspices of the Zoological Society

G. P. PUTNAM'S SONS
NEW YORK AND LONDON
The Knickerbocker Press
1926

The
Knickerbocker
Press
New York

Made in the United States of America

To

ALL THE GALLANT NAVIGATORS

from

Christopher Columbus to the Prince of Monaco

who have laid firm the foundations of oceanography

and to

the generous patrons of this work

from

Queen Isabella to Henry Whiton and Harrison Williams

PREFACE

THE origin and evolution of life, men and expeditions are interesting. On the very day of my return from the Galápagos in the *Noma,* I was introduced to a recently elected member of the Board of Managers of the New York Zoological Society, Henry D. Whiton. Mr. Whiton said to me, "You seem tremendously interested in the Galápagos; if you ever want to go back there I will furnish the steamer if you can get someone else to provide the coal." So from this generous, tentative beginning there crystallized the twenty-four hundred ton steam yacht *Arcturus,* the specified coal, a splendid oceanographic outfit, a captain and a crew, and an expedition of six months' duration, which steamed from New York to the Sargasso Sea, thence to Cocos and the Galápagos, and which secured a host of treasures, from the most microscopic beings which contribute to the surface luminescence of the sea, to a giant devilfish weighing more than a ton.

The two chief contributors to the expedition were Henry D. Whiton, who gave the *Arcturus,* and Harrison Williams who provided three-fourths of the entire cost. Other generous contributors were Marshall Field, Clarence Dillon,

Vincent Astor, the American Museum, George F. Baker, Jr., Arthur T. Newbold, Thomas S. Yates and Junius S. Morgan. Other gifts to be recorded are a sounding machine from William H. Trotter; sets of oceanographic books from Frederic C. Walcott; motion picture negatives from George Eastman; flashlights and batteries from the National Carbon Co.; a powerful radio set from the Stromberg-Carlson Mfg. Co. and the launch *Pawnee* from Harry Payne Bingham. To Ernest Lester Jones, Chief of the Coast and Geodetic Survey, I am obliged for a host of kindnesses and the loan of valuable instruments, and to the U. S. Fisheries Bureau for the *Albatross* launch and much valuable gear.

The entire responsibility for the sea-going condition of the *Arcturus,* her complete overhauling and the supervision of the building of laboratories, dark-rooms, refrigerators and oceanographic apparatus was assumed by Mr. J. R. Gordon and the naval architect, Edwin C. Bennett. Capt. Yates acted throughout for Mr. Williams, and it is to the whole-hearted enthusiasm and interest of these gentlemen that the smoothness of operation and general success of the mechanical basis of the expedition was due.

For Captain Howes and First Mate McLaughlin I have nothing but single-minded praise. No more willing, patient and capable seamen ever existed.

Several guests of honor joined the expedition for more or less brief stages, among them being

Prof. Henry Fairfield Osborn, Mr. Gregory Bateson, Mr. Herbert Satterlee and Miss Mabel Satterlee, Mrs. George Putnam and her son David, and Miss Margaret McElroy.

The scientific staff was of my own choosing, each of the seventeen members having a definite field of work, which they filled to the full extent of their ability. Without their loyalty, constant enthusiasm and coöperation, nothing of success could have been achieved.

The scientific working personnel was as follows: William Beebe, Director; W. K. Gregory, Associate in Vertebrates; L. Segal, Associate in Special Problems; C. J. Fish, Associate in Diatoms and Crustacea; John Tee-Van, General Assistant; William H. Merriam, Assistant in Field Work; Isabel Cooper and Helen Tee-Van, Scientific Artists; Ruth Rose, Historian and Technicist; M. D. Fish, Assistant in Larval Fish; Elizabeth Trotter, Assistant in Fish Problems; Dwight Franklin, Assistant in Fish Preparation; Jay F. W. Pierson, Assistant in Macroplankton; Don Dickerman, Assistant artist; E. B. Schoedsack, Assistant in Photography; Serge Chetyrkin, Preparateur; D. W. Cady, Surgeon.

The interest taken in the expedition was astonishingly deep and wide-spread, and the publicity was accurate and dignified. An unexpected result was the desire it aroused among several gentlemen to carry on oceanographic work in the same part of the world. Zane Grey visited the Galápagos in his three-masted schooner *Fisherman* after a conference concerning the possibilities of big-game fishing, and since then Harry Payne Bingham in his yacht *Pawnee,* and William K. Vanderbilt in the *Ara* have done most excellent oceanographic work, the former in the Caribbean and the latter in Galápagos waters. As late as March, 1926, Mr. Vanderbilt reported the Albemarle volcano as still in eruption.

The *Arcturus* Oceanographic Expedition, the ninth expedition of the Zoological Society, sailed from Brooklyn on February 11th, 1925, and returned to New York on July 30th. In the interval we steamed a distance of over 13,600 miles, touching at Norfolk, Bermuda, Panama, Cocos Island and the Galápagos. We brought back 11,000 feet of splendid motion picture film taken by E. B. Schoedsack, besides hundreds of colored plates and photographs. We established one hundred and thirteen stations, made hundreds of hauls with nets and dredges, threw overboard two thousand drift bottles containing data as to our identification, the date, latitude and longitude.

My object in this volume differs in no respect from that of the account of my last expedition, *Galápagos: World's End,*—a scientifically accurate, popular presentation of the high lights and vivid experiences of the expedition. As yet there has been time and opportunity for the careful identification of only a few of the many thousands of specimens collected, so that in some instances technical names are lacking in this volume. Whenever identification has been possible I have included it in the list of scientific names in Appendix B. Appendix A consists of a résumé of the bird life of Osborn Island. Beyond a final narrative chapter I have attempted no definite chronological journal.

All the details of operation, explanation of the apparatus, technical descriptions of specimens and of our individual problems, will be published in Volume VIII of *Zoologica,* the Zoological Society's scientific publication.

Chapters VI and X are wholly the work of Ruth Rose and in her rôle of Staff Historian she has collaborated with me in Chapters I, IX and XVI. Colored Plates IV and VI, and the book lining are the work of my Staff Artist, Isabel Cooper; Plate III is by Don Dickerman; Plate VIII by Dwight Franklin, while Plates I, II, V and VII are by Helen Tee-Van. In regard to the black and white illustrations, Figures 1, 4, 8, 49, 50, 59 and 60 are from drawings by Dwight Franklin; 31, 47 and 58 by Isabel Cooper; 36 by Don Dickerman; 12 and 18 by Charles Livingston Bull by permission of the Curtis Publishing Company; 2, 16, 56, 61 and 66 by John Tee-Van; 69 by Elwin R. Sanborn from the New York Aquarium, and 10 from the American Museum. All the remaining photographs are by Ernest Schoedsack.

OBJECTS AND ACCOMPLISHMENTS OF THE EXPEDITION

The avowed objects of the *Arcturus* Expedition were the investigation of the Sargasso Sea and the

Humboldt Current. Owing to continual storms the former was in such a disintegrated condition that I soon decided to postpone detailed study until a more favorable time. In the Pacific, to our surprise, we found that there was absolutely no trace of the Humboldt Current about the Galápagos. The inexplicable absence of this great, cold, Antarctic current was more than made up for by the presence of equally unexpected natural conditions.

Among the totally unexpected and inestimably valuable phenomena—the high lights of the expedition—were the great volcanic eruption on Albemarle (Chapter V); the albatross rookery on Hood (Chapter IV); the remarkable results of hundreds of dives in a copper helmet and bathing suit (Chapters III, VII, IX, XI and XII); the discovery in New York of a dramatic personage who had sought pirate treasure on Cocos for two decades (Chapter X); the temporary current rip in mid-ocean (Chapter II); and the deep sea work in the submerged Hudson Gorge, only one hundred miles from New York City (Chapter XV).

Finally, the accomplishment which, scientifically, proved the most valuable of all, was the result of my decision to make a ten-day stay in one spot in mid-ocean, Station 74 (Chapters XIII and XIV), where continual dredging yielded very remarkable collections of fish and crustacea, equivalent to any two months of the less intensive work. In fact the crustacea taken at Station 74 equal 80% of all the rest which we took in the Pacific.

CONTENTS

CONTENTS

ILLUSTRATIONS IN COLOR

xiii

ILLUSTRATIONS

IN BLACK AND WHITE

ILLUSTRATIONS

THE *ARCTURUS* ADVENTURE

1

THE *ARCTURUS* ADVENTURE

CHAPTER I

SARGASSO WEEDS AND WAVES

BY WILLIAM BEEBE AND RUTH ROSE

MOST amazingly I am floating in midspace beneath a dense grape arbor with the sun shining through a mat of yellow-green leaves and the unripe fruit glowing like myriads of jade beads. Then the air becomes chokingly oppressive—I gasp—kick out violently with my feet and shoot up through the tangled mass of olive growth. Dripping like Neptune, wreathed like Bacchus, my head breaks water in mid-ocean in a mass of sargassum weed—a thousand miles from land. Nothing is in sight except the sliding hillside of an appallingly steep but smooth swell bearing down upon me, until I shake the water from my eyes, brush aside the dangling strands and, twisting about, behold the huge bulk of the *Arcturus* silently lifting and settling a few dozen yards away. This is my first fish-eye-view of the Sargasso Sea, on the only day for weeks which is calm enough for a swim.

The thought of a grape arbor as seen from below is more than a simile of these hanging gardens, and far from original, for about three centuries ago a Portuguese spoke of them as *salgazo* or "little grapes."

While the sargassum may be falsely reported
to have been the weed that clogged a thousand
ships, yet it undoubtedly played a most important part in the discovery of America. Mutiny
among the crews of Columbus was too much of
a menace for the comforting daily sight of drifting vegetation not to be a very real mental
anodyne.

"They were astonied" writes an old translator
of Columbus' journal, "when they saw the sea, in
a manner, covered with green and yellow weeds,
which seemd to have been lately washed away from
some rock or island. This phenomenon gave them
reason to conclude that they were near some land,
especially as they perceived a live crab floating
among the weeds." And a week later they saw
"a tropicbird and such a quantity of weeds as
alarmed the crew who began to fear that their
course would be impeded."

When rumor and legend and travellers' tales
need renewed basis of fact they always turn again
to the Sargasso Sea. The supposed graveyard of
ships has ever been the incubator of fancies. The
great heart of the Atlantic has been credited with
powers which make of it almost a sentient monster,—it can draw to it ships and men, can hold
them indefinitely, spew them forth, or pull them
down to black, soul-crushing depths. Its vegetation is as dense as baled hay and has the holding
power of an octopus tentacle!

It is a terrible thing to me to destroy beliefs and
legends. Knowing however, that there were no

Fig. 1.—The Little Sea Devil of the *Arcturus*.
Diabolidium arcturi Beebe.
Three times natural size.

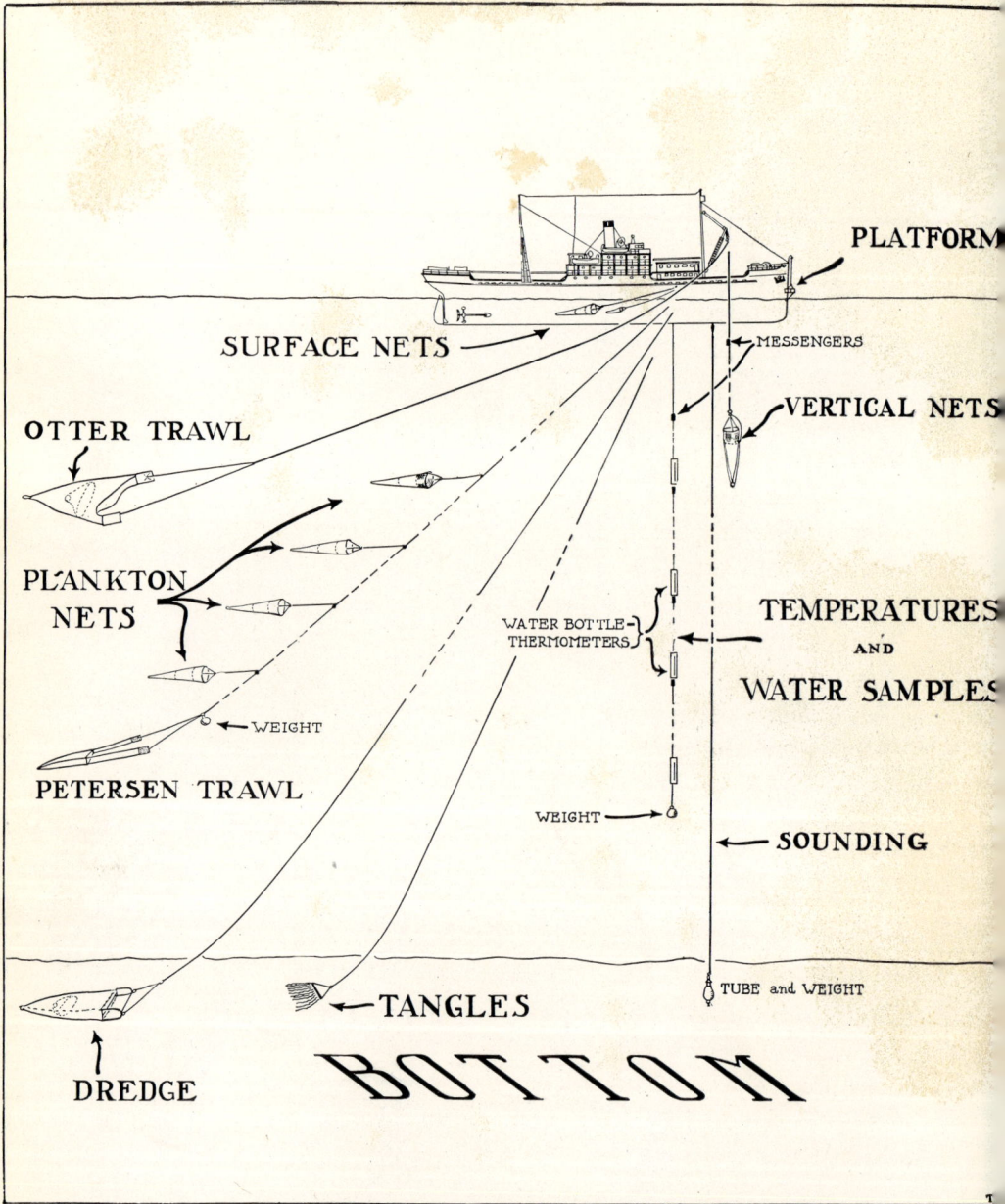

Fig. 2.—Diagrammatic View of the Oceanographic Apparatus of the *Arcturus*.
Nets, trawls, dredges, etc., used in studying the life of the sea.

fleets of vessels held captive by the sea of weed, I had nothing to abjure when I found that the only wrecks were dissolute Welsh colliers wallowing past on their unpainted way.

The mere mention of the Sargasso Sea in the list of my intended objectives was enough to inspire a whole crop of colored Sunday supplements of ancient weed-clogged vessels. As a matter of fact, realizing that scores of sailing vessels and steamers had traversed this sea again and again, and that the fauna of the weed itself was as limited as it was interesting, my object in this area was quite definite and unique. On my numerous trips from New York to British Guiana I had now and then seen, tantalizingly near, weed of considerable extent, sometimes one or two acres matted together —a golden-yellow undulating meadow. All that I asked of the Sargasso Sea was a duplication of such a meadow which I had seen more than once in areas well outside the conventionally mapped area of weeds. I hoped that the shallow and mid-sea life beneath, ranging from 100 to 500 fathoms deep, fed by the untold myriads of dead creatures falling slowly from the weed through the water, would yield hauls of unexcelled richness.

In February I took the *Arcturus* from Bermuda southeastward straight through the heart of this sea, then east, almost to its furthermost limits. Months later on our return I again steamed through a great section, this time farther to the north. We saw numberless patches of weed, but seldom any which were larger than a man's head. For many

days, in storm and calm, these averaged one to
every square hundred yards. So my Sargasso Sea
failed, in the aggregate, to materialize and I contented myself with the thousand and one other in-
terests and problems which always rush in to fill
such a vacuum.

When, in the very heart of the sea, I found only
small heads or mats of weed I should have been
truly desolate were it not that the explanation was
of exceedingly great interest. It made the Sargasso Sea more familiar, less sinister; it showed
that even in this shifting, plastic, nomadic, open-
work island there were definite seasons. An eternal
spring or autumn or winter is a frightful thing to
contemplate, while a succession of seasons links the
very antipodes with our home backyard. Dunsany
well knew how to destroy an alien feeling—to connect the extremes of geologic ages—when he began a tale with the sentence, "It was a cold winter's
evening late in the Stone Age."

And so this region lost much of its inimical character when I realized that I could say of my visit,
"It was a late autumn day in the Sargasso Sea."
My experience demonstrated an incisive difference
between an undertaking dealing with business, religion or politics on one hand, and science on the
other. We had set out to find vast fields of the
weed teeming with living creatures, and we found
only small mats and plaques almost destitute of
life. A negative result such as this would be accounted a failure in business or hopeless in religion.
To science it was of concrete value and added a

wholly new interest to the entire problem. Simultaneously with the disappointment at not seeing the fragments of weeds united into vast fields, came the certainty that following this autumn and winter, there must come spring and summer to these sunken meadows.

Although life was at lowest ebb yet the sargassum itself was in full growth. Day by day as we steamed eastward the weed became fresher and cleaner. The dark-colored, older portions disappeared in the heart of the new branches. Each bunch sent tiny sprigs up into the air, a valiant effort on the part of a poor, aquatic relation to share the thinner medium with the forests and flowers and fruits of the dry earth.

The origin and maintenance of the Sargasso Sea is still a moot question, whether the vast area is replenished annually by fragments storm-torn from the rocks of shallow coastal waters and poured forth by the Gulf Stream, or whether the weed perpetuates itself by continuous growth. Like the familiar banana, there are no seeds or spores formed in mid-ocean, but the growth of new, pale-yellow fronds and bladders is vigorous and constant. After my experience on this expedition I have no doubt whatever that the weed can propagate itself, vegetating, for a great many years if not perennially. When I kept masses of it in running water in aquariums, the older portions soon died from some excess or lack of light or heat. When I picked these pieces up by the newly sprouted fresh tips they would break off by their own weight, the old growth sinking to the bottom, while the newer sprigs and bladders rose and floated buoyantly at the surface. This would seem to account for the great abundance of wholly new heads which I observed in the heart of the area, quite devoid of any down-pulling old growths. I have no doubt that in a vast number of cases the sprouts are automatically detached at the point of juncture, either by the turbulence of the waves or after the whole has been pulled under water for some distance. That there is a certain amount of constant replenishment from coastal plants there is no doubt, but I think this is of minor importance in the maintenance of the Sargasso Sea as a whole.

The great age of the Sargasso Sea is attested by the specially adapted organisms,—fish, crustaceans, worms, anemones—which inhabit it, while the extreme reluctance of these to leave the shelter of even a tiny frond is a powerful argument against any wholesale, rapid, annual replacement of the oceanic weed-drift by fresh supplies from shore. Although we seemed to have arrived in the winter of the sargassum fauna, yet we collected 95 percent of the known crustacea and other groups in proportion.

An unexpected coincidence is infinitely more exciting and interesting than the fulfillment of a preconceived plan: Hence my delight at discovering that my most interesting days in the Sargasso Sea occurred at the same spot in mid-ocean as the most dramatic points of Columbus' first voyage.

My mind went back to the details of that expedition and as the sublime may be compared with the ridiculous, so I compare the efforts of Columbus with my own. How absurd and petty became the few delays and disappointments of my preparations when I recalled the years and years during which he passed from country to country, trying to make his convictions real, his ideals practical to one sovereign after another. The entire cost of outfitting the three caravels of Columbus was $7,-203.73 but, while this seems like an astonishingly small sum, we must remember that the purchasing power of coin at the end of the fifteenth century, for ships, labor and food, was at least twelve times what it is today. Hence it is probable that seven thousand dollars in 1492 would equal eighty thousand today. However, we must agree with Thatcher that "Under any circumstances, whether we consider the maravedis expended or the results achieved, we may regard it as the most fortunate outlay of money since gold and silver and copper were minted into coin."

Here was I with my one vessel, on an expedition which was to cost more than twenty times Columbus' original outlay, with hope of results, which even at the maximum, could be considered only as a burlesque upon his achievement. And as a final commentary let us recall that, as a result of his being the first individual on his own expedition to detect the certainty of western land, he was rewarded by the munificent annual grant of ten thousand maravedis, or sixty-one dollars, a perquisite or tip derived from the profits of the slaughter houses of Seville!

Perhaps the best way of visualizing our life and adventures in the Sargasso Sea will be to compare them with the notes of Columbus made upon geographically identical days of his memorable voyage.

> SEPTEMBER 23RD, 1492, CHRISTOPHER COL-UMBUS WROTE IN HIS JOURNAL, "WE SAW A TROPICBIRD."

On the 28th of February, 1925, on the *Arcturus,* so exactly at Columbus' location, that four hundred and thirty-three years before we would probably have been within sight of the *Santa Maria,* the dismal bleat of a tin fish-horn woke me. I slid out of my berth on the next roll of the ship and groped sleepily for dressing-gown and slippers that were arranged in fireman fashion ready to be donned. The captain and officers of the *Arcturus* had been instructed to blow the horn whenever they sighted any living thing, and their eyes were uncommonly sharp.

Like figures in a demented Swiss weather indicator, the port cabins simultaneously decanted a row of bathrobed observers, who peered earnestly at heaving grey sea and lowering grey sky. It was that dreary hour of dawn when the whole world is grey, without a hint of the color-bringing sun. Shivering, we lined up at the rail in time to see a snow-white tropicbird skim past the wireless structure, his two elongated tail-feathers trailing like a foamy wake.

There was a notable lack of enthusiasm at this sight. It was the fifth morning that we had leapt out at dawn to see the same bird, who appeared to take delight in coming only at this hour. The spectacle had somewhat lost its charm, not to speak of its novelty, but the officer on duty, like one of the immortal Six Hundred, refused to reason why. I also suspect that he enjoyed making us get up.

As we turned back to our cabins, one morose scientist was heard to mutter, "It may be bad luck to shoot an albatross, but I'd like to take a chance on that tropicbird."

Life is strenuous on an oceanic expedition, and on the previous evening our nets had brought in a rich haul of fishes from the depths, and my late session in the laboratory had lasted until three. So now I lazily determined on another half-hour in bed. Just as I was comfortably dozing off, there was a scrambling at the open porthole, and with a thud Chiriqui dropped on my chest.

Chiriqui is the small Panamanian monkey who has been the indispensable mascot of three expeditions of the New York Zoological Society. He is a much travelled and thoroughly spoiled person, and in his more destructive moments is known as Rasputin, the Demon Monk. His uncanny ability to escape from confinement causes him to be referred to also as Houdini, and the exercise of this talent accounted for his presence now. Having apparently toiled all night, he had at length succeeded in breaking out of jail, as represented by his enormous cage on the forward deck, and grinning with fiendish delight he turned three somersaults on my prostrate form, concluding the performance by scaling the life-preserver rack and from that eminence hurling himself at my head.

Feeling gloomily Shakespearean, I informed him that he had murdered sleep, and hurriedly dressed, with a wary eye on him and the more perishable articles in the cabin. He is living proof of the prestidigitator's boast that the hand is quicker than the eye, and with three snatches he can irremediably wreck as many objects. To the accompaniment of his protesting shrieks, I returned him to his prison, repaired the hole through which he had escaped, and descended to the main deck for a half-hour of work in the laboratory before breakfast.

On this, as on previous days, early morning found the ship wallowing through the endless procession of great surges which rolled tirelessly up from the south. Only thin streamers of weed, sometimes extending for a mile or two, undulated over the leaden sea. The use of the intricate deck machinery which operates our diversified gear was complicated still further by the incessant and violent motion of the vessel. We had become experts in balancing and, at a preliminary cost of a good deal of breakage, in knowing just how far we could roll before laboratory equipment suffered a sea-change into something new and strange in the way of wreckage.

Some of the scenes in the laboratory during those first stormy days defy description. An agonized scientist, caught unawares by a particularly vicious

Fig. 3.—The Boom-Walk over the Side of the *Arcturus*.

This permitted access to the undisturbed water, thirty feet away from the vessel.

Fig. 4.—Forward Hold of the *Arcturus*, where All the Apparatus for the Voyage was Stowed.

lurch, would find himself far from refuge, in the midst of a steep deck made glassy by water, alcohol, formalin and other liquids that spilled from various directions. With hands too full of precious and breakable objects to grasp at a straw, he would skate helplessly down the incline, seeming bent on dashing his head against the wall, while the rest of us, bent over desks and clinging to sliding microscopes and specimens, could throw him nothing more helpful than anxious looks and cries of encouragement.

COLUMBUS, FROM HIS POINT OF OBSERVATION ON THE HIGH POOP OF HIS CARAVEL, LOOKED OUT OVER THE SAME EXTENT OF OCEAN AS I WAS NOW WATCHING. HIS HISTORIAN TELLS US: "THE AIR WAS SOFT AND REFRESHING, AND THE ADMIRAL SAYS NOTHING WAS WANTING BUT THE SINGING OF THE NIGHTINGALE; THE SEA SMOOTH AS A RIVER. MANY WEEDS APPEARED."

This fifth morning of the tropicbird's alarm-clock appearance held the promise of being fair. Soon after sunrise the terrific swell went down and before noon the sea was the smoothest we had so far encountered. I spent an hour in the pulpit and lest any Fundamentalist be startled by the connection of pulpit and science, I hasten to explain that it is one of the several queer contraptions that make the *Arcturus* a mystery ship to passing vessels. On every sea trip I have ever taken, my favorite position is as far out on the bow as possible, looking enviously down at the floating creatures which are constantly passing. And now I had devised this pulpit which answered every requirement. It is a bit of iron grating, surrounded by a waist-high iron rail and fastened astride the bow of the *Arcturus*. It can be raised or lowered to any desired position, and this morning the weather was so promising that the supporting cables were let out to their fullest extent, so that the grating was now and then hidden by a rush of water as the ship dipped forward into a smooth billow.

The first descent of the swaying pilot-ladder was an uneasy experience, but one was not at all likely to fall when the possibility of falling was so evident,—provided, of course, that one had a "head for height." For a while I think that the captain suffered more anxiety than anyone, as from his vantage point on the bridge he witnessed the disappearance of land-lubbers over the very point of the bow. Once in the pulpit, the sensation was rather like being in Mahomet's coffin, suspended between sky and sea, with nothing under foot but a few strips of widely-spaced metal, hanging under the cliff-like bow of the forging vessel that slid down the watery slopes in a ceaseless attempt to overtake and crush me and my scant support. There was no sound but that of the rushing water cleaved and flung aside by the sharp prow. The sun's rays tapered into a luminous cone that plumbed infinite blue depths just ahead, a hypnotizing focal point for dazzled eyes. From undulating blue meadows a school of flyingfish skittered like grass-

hoppers from a hay-field, and two or three of them
skimmed knee-high across my little platform.

"They Saw" We Read in the Record of
Columbus' Voyage, "Vast Schools of Tunny-
fish, and the People on the *Nina* Killed
one. The Admiral says that those Indica-
tions Came from the West, 'Where, I Hope
in the Exalted God in Whose Hands are
all Victories, That Land Will Very Soon
Appear.' "

Standing in my pulpit, as I have said, I turned
to watch the flyingfish plump back into their briefly
deserted element, when a dark shadow shot through
the water toward me,—the tunnies had come,—and
after this hardly a day passed when from four to
forty could not be seen, swift, violet torpedoes keeping as steadily in our path as if fastened in some
inexplicable outboard manner to our keel.   But
today they did not remain long.   I happened to
be looking when the whole school turned, as one
fish, and with lightning speed darted out of sight.
The reason became apparent when seven advance
scouts of a gang of dolphins rushed up and wheeled
into line, attracted by the throb of our engines from
heaven knows what distance, to that game of which
dolphins never seem to tire.

I don't suppose there is any more inspiring sight
than a school of dolphins leaping round a ship.
They are so unmistakably and thoroughly enjoying themselves, in their effortless rush and curving,
easy leaps, that no one could help feeling that al-

most affectionate sympathy which is inspired by watching anything done superlatively well by someone who has tremendous fun in doing it. Right under my feet these friendly creatures now frolicked, so close that the lift and fall of the ship, sometimes synchronizing with their motion, made me feel that I was riding one of the powerful curved backs that slid from water to air and back again so smoothly as to throw scarcely a drop of spray. The torpedo bodies, perfectly fashioned for just this, accurately held the appointed distance from the ship and seemed not to move a muscle. Only close scrutiny revealed the terrific power of almost imperceptible strokes of the broad tail flanges.

My instant reaction to a school of dolphins is an irresistible desire to shout, and this, being the first combination of pulpit and dolphins, made me excitedly wishful for the laboratory toilers. I was too selfish to leave this delightful post in search of the rest, so I lifted up my voice in what seemed fruitless shrieks against the towering ship's side.

Presently a head peered anxiously over the rail far above me, and seeing that my cries were not for help but for appreciation, vanished for a moment and reappeared with an augmented audience. This was still in such an early stage of the cruise that the Captain suffered almost hourly pangs of apprehension by mistaking screams of enthusiasm for calls for succor. Long before the end of the six months' voyage nothing less than the wail of a banshee would have attracted his attention.

The capacity of the pulpit being three, two of

the audience joined me and for awhile we amused ourselves by trying to touch the gambolling dolphins as they shot up from the water.   For half an hour or more we timed individual dolphins with a stop watch, and found that they came up for breath on an average of once every three minutes, the inhalation through the open blow-hole lasting from three-fifths to an entire second.   Once we were thoroughly soaked by the plunge of the bow into a deep trough,—a breathless moment when the actual security of our position was forgotten and the whole ocean seemed to overwhelm us.   When the dolphins tired of us and rushed away on some suddenly remembered errand, I mounted to the deck and lowered to the two who remained below the long-handled net with which specimens were scooped from the waves.

Pieces of weed were constantly passing, each one with its assortment of little beings who depended upon it for protection and whose lives were bounded by its fragile shelter.   Sitting astride the bulwark, I hauled up a bucket full of weed, lowered an empty one into its place, and carried the catch down to the main deck, where it was put in a tub and carefully examined for its inhabitants.   I remembered that I was not the first collector in this identical spot, since four centuries earlier a famous explorer had proved himself a worthy carcinologist;

"At Dawn They Saw Many More Weeds, Apparently River Weeds, and Among Them a Live Crab, Which the Admiral Kept" (Columbus' Diary).

From my own modern bucket quaint things came forth,—innumerable tiny crabs and shrimps, perfectly disguised in the yellow-brown colors of the weed and even reproducing on their carapaces the shapes and tinges of the blemishes and parasites on their vegetable home; absurdly attenuated pipefish, hardly to be detected when in motion, so exactly could they imitate the undulation of a waving frond; naked mollusks, or Nudibranchs, incredible creatures that must be seen to be believed and cannot be described; infinitesimal worms and snails, furnishing food for larger forms and themselves finding some microscopic fodder in their watery jungle; and each species wrapping itself in a cloak of invisibility and melting into its background with magical completeness. The commonest crab was undoubtedly that which Columbus collected, and which bears the name of *Planes minutus*.

On the scattered bits of sargassum which we salvaged, I found many hints of the spring which was to come to this strange land of sea tares. Masses of snail eggs,—some in many-celled stages, like diminutive parodies of golf-balls, others with active embryos pushing and straining to break through the membranes and begin that series of hopes and fears which both snails and we call life. Now and then were skeins of fish eggs tangled inextricably among the fronds,—linear nurseries of thousands of brothers and sisters.

I took a little three-inch frond of weed into the laboratory and watched it under my binocular microscope. I pretended the common little inhabi-

tants were rare and began to observe instead of
merely to see them.   There were three kinds of
hydroids,—the palms, the trees and the spiked
clubs, all superlatively dainty and elegant, foresting
these diminutive roof gardens of the sea.   Here and
there were more formal plantings, row upon row of
beautiful ivory or alabaster chalices, from which
sprang severe fountains of tentacles,—minute bryo-
zoans or moss animals,—all arranged just so, like
the alabaster vases of Italian gardens.   The bryozoan beds were still exquisite when their occupants
were dead and gone.   The sere and autumn of the
moss animals' year left a mosaic of thousands of
flattened hexagons as perfect as honeycombs, as
translucent as age-old moonstone.

These serried ranks of the bryozoan folk are all
flattened against their world of weed, but the wavering groves of slender hydroids are connected at
their base by rootlets or stolons which wander and
weave about the fronds and the bladders.   It is
hard to say what are the relationships of a
group of these little hydroid palms.   Is the
tall animal flower at the summit of the berry-
like float the child or parent of the one behind, and these three which stand up tall in a
row like the masts of a Lilliputian wireless station,
—are they cousins or brothers?   If however, we are
confused at this relationship, what can we say of
the actual transition from one generation to another,—as astounding as it would be for a cat to
have geraniums instead of kittens, and the plant
offspring to scatter puppies in place of seeds!

The discovery of our first specimen of *Ptero-phryne* drew everyone's attention; the youngest member of the staff took one look at the little creature and cried in honest ecstasy, "My Word!" and so it was christened on the spot, and so, during its brief span in our midst, it was affectionately called. I do not blame anyone for objecting to the adjective "fascinating" as applied to a fish, but I ask such a sceptic to wait until he has seen *Pterophryne,* the Sargasso Fish *par excellence.* From snout to tail-fin it was the piscine essence of the fronds, its fin rays produced into finger-like appendages, with which it crept about in the weed, swinging from frond to frond, dangling upside-down, and assuming postures that were irresistibly comic. Its foolish face was fixed in an expression of intense earnestness, and the stout little body performed amazing antics with the agility of a monkey. I hold no brief for fish as pets, but *Pterophryne* is the exception. Everyone who could draw clamored to paint this specimen, others inspected it with a view to determining the species, and some of us wished merely to watch it and chuckle. Soon the ghastly blue of Cooper-Hewitt lights issuing from the bridge-casing told that moving picture and still cameras were busily recording its appearance and activity. In my journal I find a sad note for the evening of that date; " 'My Word' died of publicity." He had his crowded hour.

Elsewhere I shall describe more in detail the various forms of apparatus used on the *Arcturus,*

FIG. 5.—The Large Dredge Being Hauled in After Dragging Along the Bottom for Two Hours, at a Depth of One Mile.

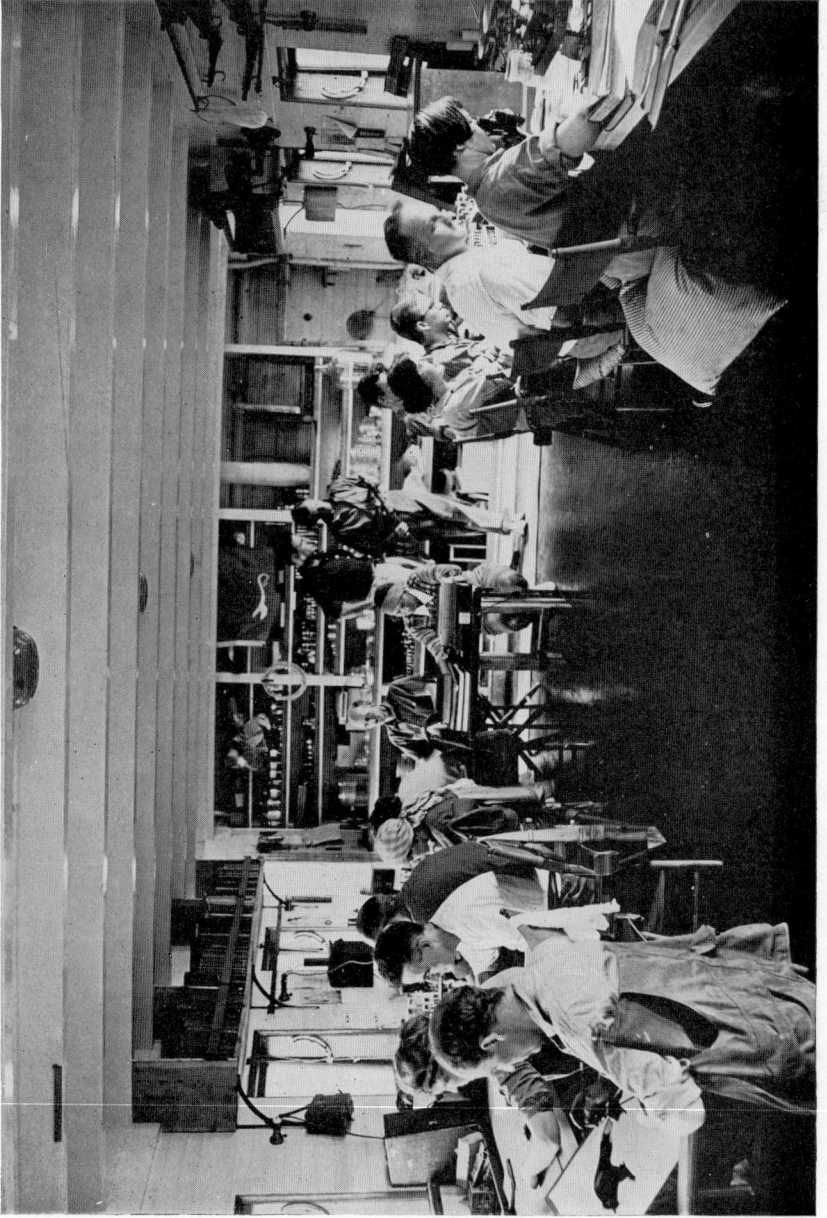

Fig. 6.—Scientific Staff of the *Arcturus* at Work in the Laboratory.

but I must mention the boom-walk before going on to the deep-sea part of our typical Sargasso day. When I saw the *Arcturus* in dry-dock the thought came to me how much of a vessel is outside and how little anyone has ever made use of it. I remembered Howard Pyle's drawing of a pirate's captive walking the plank and I made up my mind to adapt this to the uses of an oceanographic expedition. I fashioned two thirty-foot booms rigged outboard on the port side, one slightly above the other and about three feet apart. To these, by a many-looped rope I laced a duckboard walk. When swung at right angles to the vessel's side and firmly guyed, I had a perfectly safe runway extending far out from the ship and over quiet water, beyond the foaming wave thrown up by the passage of the *Arcturus* on her course. I could walk out in calm or in storm and, from a curious, semi-detached view-point, contemplate the ship plunging through the water. One was of the vessel, and yet not exactly in it nor on it, a state of mind which may resemble that of a soul in its astral body looking back upon its corporeal one. Searching for a name that should express the feeling of this position, we hit upon the Fourth Dimension as most appropriate.

In the trough of a swell which looked inconsiderable but felt mighty, the tip of the boom described a great arc, swinging far up into the air until one looked down from an appalling height on the main deck, then swooping waveward with such velocity that a salt bath seemed inevitable. It was a glorious experience if one was a good sailor.

The uses of the boom-walk proved to be manifold, so much so, indeed, that the Captain came to me one day to apologize for the scepticism which he had shared with marine engineers and others in the ship-yard where the weird contraption had been made and attached. Trailing silk surface nets from the extreme end of the boom-walk proved infinitely more effective than the conventional method of trailing them over the stern in the roiled and disturbed wake. When we were anchored I was able to use the outboard walk as an auxiliary boat boom, or a place from which I could make a descent in my diving helmet even at night. The sounding davit was fixed half way out and we trolled for and harpooned dolphin fish and sharks, besides using it for photography, and catching up weed, fish and organisms of all kind. In fact this, together with the pulpit, increased our totality of effectiveness to an astonishing degree.

Among the host of creatures which we took in quarts of plankton in our surface nets by day and by night in the Sargasso Sea, one is especially worthy of mention in this place.

Leptocephalus is a general, ignorance-confessing name given to the larval form of eels (Fig. 10). My first introduction was when I looked at a small aquarium of plankton and saw a half dozen mother-of-pearl eyes swimming around quite by themselves. This was after I had been studying plankton for a few dozen hauls, and had passed the stage of wondering whether excess of microscopic work was working injury to my eyesight. Yet even now I did not quite believe what I saw, until I dipped in my hand and lifted out a twelve-inch piece of flexible water. There was absolutely no structure to be seen except the gleaming eyes, and yet here was a living fish. When dead and preserved, the body, shaped like a long thin willow leaf, became translucent and then it was possible to make out the hundred-odd delicate segments and the all but invisible gills and stomach. When the head was placed under the microscope there leaped into view a regular old-fashioned dragon, with enormously long, sabre teeth, which, were the animal twelve feet instead of twelve inches in length, would make it infinitely more dangerous than the largest anaconda. In the Sargasso Sea we took hundreds of specimens of many species, only a very few of which can be accurately identified, for the reason that we lack the connecting stages between these indefinite water wafers of organisms, and the more palpable adult fish.

The history of two forms of *Leptocephalus* has only very recently been worked out, and is another of the inexplicable complexities of nature, which to our practical, human minds seems an absolute waste of energy. To Dr. Johs. Schmidt belongs much of the credit for the patient unravelling of this astounding problem. As *Leptocephalus* is strange as any dragon in a fairy tale, so its life history equals the unreality of any fairy tale itself.

Briefly, these watery beings which, at night, we captured in dozens in our surface nets, are hatched from eggs which are deposited not far south of my first Sargasso objective, 30° North and 60° West. At least two species of these tiny, new-born *Leptocephali* soon begin to swim slowly northward, reaching the latitude of Bermuda within the first year. They then separate into two mighty streams. The one which swings westward develops rather rapidly and soon after the first year has changed into young eels or elvers, and, guided by some instinct to which we have not the slightest clue, seeks the various fresh-water streams and rivers from Florida to Canada from which a year or more before, their parent eels emerged.

The offspring of European eels, on the contrary, turn to the east and take three years to reach the mouths of their ancestral rivers—be they British, Spanish, French or Norwegian. Here they wriggle slowly up the saltless currents, and after a dozen years or so, play their part in this marvellously intricate round of life. In a single haul of a metre net at 30° North and 60° West it is possible that we captured two *Leptocephali,*—one of which would have completed its growth in the farthest tributary of Lake Ontario, and the other in some little stream of the headwaters of the Rhine.

Why should such sedentary creatures, spending almost all their lives in a single reach of brook or stream, suddenly be moved to traverse thousands of miles of open ocean, braving voracious fish and cetaceans to lay their eggs in the Sargasso Sea close to an alien continent, when others of their class successfully spawn under the nearest pebble?

Is there a more dramatic phenomenom in the world than a whole generation of adult eels of two continents moving majestically in their millions,—setting out upon a voyage at the end of which each female will scatter her ten or more million eggs, and from which no eel will ever return! When, within a space of several years, learned ichthyologists wrote confidently of eels descending to salt water and, inside of a month, depositing their eggs close to shore, we can hardly afford to laugh at Aristotle who, two score centuries ago, stated that eels have no sexes, nor eggs, nor semen, and that they rise from the entrails of the sea.

So far on this day we had concerned ourselves only with the surface life of the ocean, but now we prepared for some deep sea work. The preliminary was to take soundings to determine how far down the large trawl could be lowered without scraping bottom. The warning word ran round the ship, "Watch your desks, she's going to roll." Of course we had to stop in order to sound, and everyone dreaded it, for it meant that the *Arcturus* would soon swing into the trough of the sea, and that everything not bolted, wedged, reinforced and clamped would take unto itself roller-skates or wings, or achieve the same effect. Talk about the origin of life upon the earth! no day passed without a score of examples, in full speed mutation time, of spontaneous generation, of metamorphoses of ink bottles, jars and filing boxes, into sepia lakes in which swam long preserved fish and over which fluttered innumerable snowflakes of catalogue

cards. In those days in the Sargasso Sea, that tried
men's souls, as well as more material portions of
their anatomy, we endeavored to accommodate ourselves to the whims of the ocean by voyaging as
much as possible into a head-sea. Thus we only
pitched, not nearly so distressing and violent a motion as rolling. If the ship fell off a bit, or it was
necessary to change the course and an unexpected
roll disturbed the laboratory toilers, there was never
lacking some one to dart out and cast a black look
toward the bridge, as one who would say "How
dare you let this ship roll!" I suppose that this as-
sumption of perfect control on the part of the Captain was really very flattering, if we could have
made him see it in that way.

The engine room telegraph clanged around to
"Stop!", the bulky iron weight and hollow sounding tube were fixed on the slender piano-wire and
the humming descent to the depths commenced.
So did the rolling. The boom-walk was already
occupied by one man watching to see that the wire
did not kink and another carefully taking the angle
at which the wire entered the water.

The indicator-arm of the sounding machine at
last jerked sharply downward as a signal that the
weight had touched bottom and detached itself; a
brief pause and the little motor began to whirr
again, reeling in a mile and a half of wire, which,
as it came, was wiped and greased before it reached
the drum on which it was recoiled. Every time
a sounding was taken a weight was abandoned on
the bottom, and considering the number of sound-

ings that have been taken in years past, one's imagination pictures the ocean floor as thickly and bumpily carpeted with seventy-five-pound pear-shaped balls of iron. The cold light of statistics, however, reveals the fact that so little is actually known of the depths of the ocean that, outside the thousand-fathom line, there is in the Atlantic an average of only a single sounding record for each twelve thousand square miles. So, after all, the ocean bottom is far from being cobbled with iron.

When the sounding tube broke the surface on its return journey, and was emptied of the sample of the bottom which it had sucked up, any absurd fancy about man's puny efforts was banished. The dishful of Globigerina ooze was a pinch of the stuff with which millions of square miles of the submarine world are covered. Under the microscope the greyish white gravel resolved into the fragile shells of infinitesimal creatures, which in unthinkable quadrillions spend their lives floating near the surface and, dying, sink slowly through the black depths to add their tiny homes to the vast piles of their fellows'. In a world without color, because it is without light, totally lacking in vegetable life, where an unchanging iciness of temperature prevails, and where the pressure to the square inch amounts to an added ton for every added mile of depth, there are huge areas where the bottom is deeply covered over by the bleached remnants of these single-celled little beings, each smaller than a grain of sand. And over them swim and crawl and grope forms of life that are too strange to be credited.

With the hope of getting some of these grotesque creatures of the deep, the big trawl was let over the side, and the cable began to run off the huge drum, passing through a succession of blocks that made it look as though a gigantic game of cat's-cradle was in progress on the forward deck, before it ran over the tip of the outswung boom and down into the water. At intervals of a hundred fathoms the unwinding process was checked long enough to attach a fine silk net to the cable, so that the various levels of the sea would be combed. We were once more under way, going at slowest speed—about two knots—so that too great a strain might not be put on nets and machinery, and though the ship rolled a bit now and then, it was no longer the catastrophic wallowing that made us long to be limpets. It was necessary to let out the cable slowly, as we had learned by experience. On one occasion when impatience overcame discretion, yells of horror greeted the sudden rising from the waves of a Gargantuan tangle, the result of too swift a descent that had allowed the cable to overtake itself in loops and coils and ingenious Gordian knots. The steam winch was checked only just in time to prevent the whole mass from striking the first block and working tremendous damage.

With the trawl at a depth of a mile, and five silk nets trailing at hundred-fathom intervals, we steamed slowly along for two hours. Deep-sea

FIG. 7.—Harpooning a Dolphin.   FIG. 8.—Catching surface specimens.

FIG. 9.—Forward view of the pulpit.

THREE VIEWS OF THE BOW PULPIT.

FIG. 10.—THE BUBBLE GLOBIGERINA.

One-celled animals living near the surface, whose shells go to form the Globigerina ooze which covers thousands of miles of sea bottom.

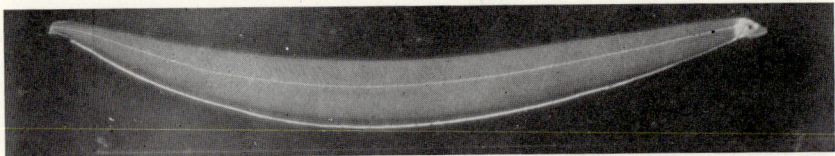

FIG. 11.—LEPTOCEPHALUS.

The flat, transparent larval stage through which the young of all eels pass.

trawling is like an enormous—and expensive—grab-bag; after all the time and labor involved in putting over and bringing in the apparatus, the sum total of the effort may be nothing at all, or it may be a host of beings strange and rare, or absolutely new. The oceanographer can trust only to luck—aided somewhat, of course, by a knowledge of the sort of ocean bottom over which life is most likely to be abundant, and in some localities, by the experience of his predecessors in the work.

Finally the shout was heard, "Beaters wanted!" This sounded like an advertisement by the owner of a pheasant preserve, but was really the result of finding that the best way to dry the incoming cable was to knock off the water with heavy sticks. Two at a time, we took fifteen-minute turns in earnestly belaboring the big steel rope before it reached the drum on its return journey. At this moment listen, if you please, to the sounds on the deck of the *Arcturus:* The staccato whacks of the beaters, pounding in rhythm to the chanty of some ballad of old England, learned from our negro paddlers in Guiana jungles; this mingled with the rumble and clank of the winch; in the laboratory typewriters clattered, the Van Slyke machine operated by the chemist thudded swiftly; photographic lights fizzed and spluttered in the bridge-casing; the second mate, sacrificing his watch below, mended nets on the whirring electric sewing-machine; while over the mechanical uproar of the *Arcturus* sounded the shrill chatters and yelps that told of an argument between Chiriqui and the ship's puppy—a canine of mysterious pedigree and unknown breed.

Descending into the forward hold for a fresh supply of vials, I delved among cases in the swaying shadows; here in the lowest depths of this wooden ship new noises drowned out everything that was happening above. The huge curving timbers of the framework might have been those of an old galleon, and the gobbets of red paint that showed where the bolts were placed were shudderingly reminiscent of those dreadful significant splashes with which, in fiction at least, all pirate ships are plentifully besprinkled. From the sounds it was easy to imagine that the vessel was on the point of disintegration; such creaks and groans must herald disaster, and when a large swell came, the grind of straining planks, and the volley of crackling, which might have come from a machine-gun nest, were deafening.

The first net was in; the winch stopped while it was detached and brought over the side and its contents gingerly emptied into glass dishes and bowls; four such nets were safely recovered, but something had gone wrong with the fifth, that which had been down to five hundred fathoms; the light ropes at its mouth had twisted, evidently on the way down, for it was wound up in such a way that nothing had been able to enter it. A certain percentage of such accidents is to be expected, but as the voyage went on, mishaps of this kind were rare.

The arrival of the big trawl was the signal for

a rush hour on the forward deck; everyone, except possibly the stokers and the officer on watch, crowded around to see the catch. After the first week the crew was convinced of our insanity. Their standard of excitement was governed entirely by size, and to see fourteen grown-up people go into ecstasies over such tiny specimens was to them one of the funniest and most inexplicable sights in the world. What if we did catch a fish whose eyes stood out on stalks almost half as long as its entire body, and through whose transparent skin a minute heart and nervous system were plainly visible? If the whole creature was less than three inches long, the crew derived nothing from it but a hearty laugh. As the majority of deep-sea animals are small, the sailors seldom lacked comedy. On one occasion, when there was a shout of "Whales astern!" and every door erupted flying figures that raced aft, the oldest able seaman, a big, bored Scandinavian, was heard to mutter, "I seen plenty whales. I never seen such funny folks."

There were hundreds of specimens that must be sorted out as fast as possible, and soon every desk in the laboratory had an absorbed worker, armed with forceps, spoons and pipettes, disentangling fish from sagittæ, crustaceans from jellyfish, squids from siphonophores. If it were only feasible to label the nets "For fish only," or "Jellyfish enter here"; the oceanographer's life would be much simplified. The heterogeneous mass that is scrambled together by a trailing net is mostly of such fragile structure that it seems a miracle to float out a double handful in a dish of water and find that most of the animals are not damaged. It appears incredible that the contact with the net and the impact of the water on the upward journey should not crush all but the largest and toughest.

There was an excited shout from the dark-room that caused a stampede in that direction. In the nearly total blackness of that very inaccessible compartment, streaks and gleams and sparks of glowing light moved slowly and erratically about. In the babble of questions from a dozen people who were tripping over each other in the dark, I shouted out,

"Astronesthes and Oneirodes!"

This was not an ancient Grecian oath, but the names of two luminous deep-sea fishes that were nobly gratifying the hope with which they had been hurried into the dark-room. Brought up from a region where the pressure on their small bodies was hundreds of pounds to the square inch, into an unfamiliar zone where it amounted to only fifteen pounds, it was marvellous that they lived to reach the surface, to say nothing of continuing to exist long enough to show those little lights which up to this moment had been gliding about the cold blackness of the great depths.

Both of these particular ones were velvety black of skin. *Astronesthes* was rather slim and long-bodied, with a slender tentacle trailing from its chin, which, to my surprise, was delicately luminous down its entire length, only the thickened tip showing no light. This very fish we later captured at the surface at night. *Oneirodes* was a globular little fish, chiefly mouth; from the top of its head sprouted an appendage, the upper half of which bent at right angles to the base, and from the end dangled a tiny light, for all the world like an electric bulb. This hung before the fish as it swam along and presumably attracted the small creatures upon which it fed. Approaching to examine the illumination, they would be engulfed by the gaping mouth, so ridiculously disproportionate to the size of the fish behind it. This, however, is at present pure theory.

A third common source of illumination were the fish belonging to a group known as *Myctophum*. These too are found at considerable depths, while at night we also took them at the surface, sometimes in large numbers. They are spotted all over with brilliant points of light—the sides exhibiting a pattern that varies according to the species, and the lower surface literally ablaze with a display which presumably attracts edible creatures in the same way that the little baited rod of *Oneirodes* lures food.

With every haul of the nets bringing in these and other marvels to be studied, painted, described and classified, it is no wonder that working hours lengthened insensibly, and that the necessity for sleep was but grudgingly admitted. There was too, the ever-present peril of missing something, and Argus himself might have found his equipment unequal to the task of having at least one eye always ready for emergencies. There was a discouragingly vast expanse of ocean to watch for possible excitement.

On October 8th, 1492, Columbus Tells Us "There Were Many Small Land-birds and one was Taken Which was Flying to the South-west. . . . All Night Birds Were Heard Passing."

This voyage of voyages was thus in the height of the autumn migration southward. When I steamed slowly through the same waters it was late February, too early for the spring migration, yet twice we too saw "small land-birds," once a sparrow of unknown species, and again a robin, which had been blown far from land. The sparrow rested on our deck at dusk and could not be discovered next morning. The robin circled twice, but although hundreds of miles from shore, set bravely out westward without alighting.

Another curious sight which Columbus could not have seen was a gull with jet-black breast and under parts. It so defied all my attempts at identification that I shot it as it soared high over the deck. It proved to be a kittiwake in good condition but with the ventral plumage saturated with oil, into which it must unwittingly have swam. It had fed heartily on the small shrimps and crabs which make their home among the sargassum weed.

Another Entry in the Log of Columbus: "The Sea was Very Calm, for Which Reason Many Sailors Began to Swim. They saw Many Dorados and Other Fish."

In the afternoon of the typical day of which I am writing the *Arcturus* stopped again for the purpose of giving us an opportunity of using the reversing thermometers,—ingenious instruments for obtaining temperatures and samples of water at different depths.   The waves had been smoothing out all day and finally it looked possible to take an ocean swim.   We had planned that such bathing would be a regular part of the program in this sea where, we had fondly believed, calm waters were the rule, but as none of us had experience in English Channel contests, we had so far gazed on the boisterous waves without enthusiasm.   We took instant advantage of the present comparative placidity, and a pilot ladder was unrolled over the side.   Down this, those who were unwilling or unable to make a high dive, conservatively descended, and discovered that the placidity, noticed with such satisfaction from the deck, was only comparative.

I swam rapidly away for fifty yards and then turned and gave myself up to a realization of my position in relation to old Mother Earth.   A glance around brought a tremendous thrill.   The swells were smooth but mountain high, not wave-like but as if the whole horizon were a range of mountains marching majestically toward me.   My own movement was negligible; I seemed for a long time to be floating at the bottom of a gigantic ultramarine cone, then slowly and gently to rise—high, high, higher,—until I dominated the *Arcturus* and seemed to approach the drifting clouds overhead.

Yet no matter to what height I was borne, the distant horizon always held another, still more lofty ridge.

These great swells fittingly suited the dramatic location—half-way between America and Africa, actually balanced between Florida and the Sahara. The buoyancy was unbelievable and the difference between swimming in a few feet of salt water and here where there were two or three miles of liquid beneath me, seemed very noticeable.

I dived and entered an ultramarine world, with sprigs of amber sargassum weed floating near the ceiling of that world. Tiny fish darted past, and once, even with the dullness of my aquatic vision, I saw a small school vanish from view—a group of timid flyingfish which took to wing and entered the air at sight of my strange appearance. I dived a second time and sank as low as my stored-up breath permitted, and then before I turned and kicked upward, took one long look beneath and tried to imagine that unimaginable world of life down, down in the ever blackening, ever greater pressured depths. No ship or companion was visible and my sense of devastating isolation, of cosmic awe can never again occur with equal force in this life, unless, some day I am able to sense Tomlinson's experience when,

> "A Spirit gripped him by the hair
>   and carried him far away,
> Till he heard as the roar of a rain-fed ford
>   the roar of the Milky Way."

SCARLET-TAILED TRIGGERFISH

Xanthichthys ringens (LINNÉ)

Various color phases of the same individual. The blue phase is the normal surface color, the white is assumed just before death. (Top figure natural size, the others reduced.)

SCARLET-TAILED TRIGGERFISH
Xanthichthys ringens (LINNÉ)

Various color phases of the same individual. The blue phase is the normal surface color, the white is assumed just before death. (Top figure natural size, the others reduced).

PLATE I

Here then, in the midst of the sea, for a moment, I peered down toward the mid-ocean ridge which Wagner would use to fill up a chink in his continental mosaic; which some would have as the site of old Atlantis, or others strew with the weed-caught wrecks of ancient galleys, medieval ships and modern dreadnaughts. But no theory, whether plausible or incredible, could ever people these depths with beings stranger than those piscine elves and hobgoblins which we were soon to draw up into the light and warmth of our daylight.

I followed the last stream of my life-bubbles to the surface and slowly barged along toward the *Arcturus*. From my fish-eye-view the ship looked enormous, a towering wall of white lifting to show dark, incurving expanses of slimy wood below the water-line, and then plunging down with pile-driver force as though to smash the impelling wave that shot out from the bow in splintered foam.

Getting aboard again was a nice problem in judging time and distance; to grasp the floating ladder on the downward roll and allow the reverse movement to hoist you up without scraping you along the timbers, to employ the next few seconds in climbing high enough so that the next downward roll would dangle you in mid-air instead of sprawling you into the water again; and finally to accomplish all this without losing goodly portions of skin, was a game that required practice to be well done, and luck to be done at all.

As the last swimmer slid damply over the bulwarks to the deck, the fish-horn sounded dismally from the bridge and an arm over the weather-cloth pointed abeam. We obediently gazed and suddenly two huge-flanged tails heaved up, hung quivering with giant vibrations, hit the surface almost simultaneously with mighty smacks and were gone. Whether we had glimpsed a battle, a courtship, or merely a frolic of two monster whales, we did not know.

Four hundred and thirty-three years ago almost at this very spot, the sailors of Columbus had seen many dorados, and today, at our early dinner, while sunset colors were still reflected in the all-surrounding waters, we heard shouts from the boom-walk, and fled to the deck, to find that a trailing hook had been taken by a big *Coryphaena* or dolphin-fish, or, *como se llama en Español—Dorado*. A vigilant deck-hand and the wireless operator were struggling to hoist it to the swaying, narrow boards. The gleaming fish, fighting gallantly, came out of water; the gaff lifted it over the boom, and just then the ship rolled, the dolphin gave one desperate flop and flung itself off the gaff, and the operator's feet slid out from under him. He fell face down on the steep slope of the foot-way, but under him was the dolphin and both arms were locked about it in a grip of death. We cheered him from the upper deck as he regained his feet and staggered grimly to the bulwarks with his prey. The last of the daylight shone on the green and blue and gold of the dolphin's sides, and we gathered about to admire perhaps a direct descendant of Columbus' fish. The first officer, who had been in charge of most of our deep-sea hauls, passed by. He paused, glanced at the victim, and remarking casually to no one in particular, "Well, thank God, somebody's caught a *visible* fish," he moved on down the deck.

"THOSE ON THE CARAVEL *Pinta*" SAYS CO-
LUMBUS, "SAW A REED AND A LOG, AND THEY
ALSO PICKED UP A STICK WHICH APPEARED
TO HAVE BEEN CARVED WITH AN IRON TOOL, A
PIECE OF CANE, A PLANT WHICH GROWS ON
LAND, AND A BOARD. THE CREW OF THE *Nina*
SAW OTHER SIGNS OF LAND."

Such were the signs which cheered the Great Navigator and his men and made them feel that land must be somewhere there below the everlasting western horizon. The same night, in the darkness, the vibrations from a tiny light were detected by the keen eyes of the Admiral himself—the first direct contact with the New World. When we were twelve hundred miles out in the Atlantic, close to Columbus' route, I stood one evening alone watching a new crescent moon hung upside-down in the sky, and wholly obsessed with the vastness and loneliness of the great ocean. Later I went into the library, and turning to the powerful radio which had been given me, I idly put it into commission.

Instantly there arose a confused sound of instruments which, almost at once, cleared into a full orchestra, in a concert hall in far distant Pittsburg, playing "Hands across the Sea." Another

half-inch twist and the room was filled with the
liquid tone of some unknown Señorita, singing a
song of old Madrid in a far-off Spanish cab-
aret. It was beyond words miraculous to realize
that the whole atmosphere above this mighty ocean,
so clear and silent in the moonlight of the Sargasso Sea, was vibrant with untold hosts of melo-
dies streaming past from all over the world. I shut
off the radio, went on deck far up into the bow, and
looked down into the silvered water, my eyes straining as had those of Columbus. I knew then that
all the marvel of our modern inventions, all knowledge of the restless millions of people on the distant continents could arouse no emotion equal to
his, when, four centuries ago, the first glimpse of
that tiny light came across the water.

CHAPTER II

WHY has no one ever written of walls and fences? They are full of interest, and when considered from the point of view of the fences themselves, rather than what they confine, they are very new and fertile subjects. There are invisible fences, like the miles of wire on our western plains which shine out only near sunset, until the autumn tumble-weed makes them conspicuous all day, piling up fluffy but visible barriers. The stone fences of New England seem indestructible, but when the hands that built them are quiet or have gone cityward, they drop, stone by stone, to the ground and are scattered again. But even then their paths can be traced for years by the lines of cedars and cherries, bird-planted, carried there by the wings of hundreds of generations past. There are temporary fences, like the slanting sections which appear at exposed places along railway lines to catch and drift the driving snow; and, still more evanescent, the wooden walls which are erected for the purpose of training police dogs to jump.

We in this country do not know how terrible

fences can be until we have seen the dead-fall
bamboo lines of the Bornean Dyaks, which wind
up and down hill through the jungle, and each
morning are a shambles of pitiful dead things,
from moon rats to argus pheasants. And it will
be decades before we can ever know the beauty
of English wall-trained fruit trees, planted long
before we became a nation.

It were easy to think of scores of others, but I
wish only to get my mind in the mood of thinking barriers, with all the details cast aside and only
the abstract remaining.

Walls can be more than tenuous, they can be
actually invisible, as when I once camped by the
rim of a great abyss near southern Tibet, up which
there poured so steady a wall of wind that I used
to lean recklessly far out against it, farther than
from where I could possibly recover my balance
in the event of its slacking. It was a fool stunt,
now that I look back upon it, but it showed me that
the air could offer a support like a board.

I am leading up to a wall of water, not the kind
which once banked up in the Red Sea, but one that
we came on unexpectedly in the Pacific Ocean.

On March twenty-eighth we made the transit of
the Panama Canal, and prepared to investigate
the life of that part of the Pacific which, though
on the Equator, is traversed in a northwesterly
direction by the cold Antarctic stream known as
the Humboldt Current. This is a reversal of the
conditions brought about by the Gulf Stream,
and is responsible for many paradoxical facts, such

as the presence of those Antarctic creatures, penguins, living and thriving under what should be the intense heat of the equatorial sun.

Just as in the Atlantic we had started out with the dominant idea of Sargasso Sea in mind, so now in *der Stiller Ocean* it was the Humboldt Current that we looked forward to studying. Our memory of two years ago on the *Noma* was still vivid,[1] when the turn of a promontory meant sometimes such a drop in temperature that, even while crossing the Equator, we hastily donned sweaters. A few miles made all the difference in the water, whether it flowed about our bodies comfortably warm as the tropical sun could make it, or whether it met us in our dive with the shock of a New England plunge.

The first three days in the Pacific we could think of only one thing—the glorious smoothness of the ocean. For weeks we had wallowed almost bulwark deep in the Sargasso, with never respite for efficient dredging or trawling, or a chance to walk steadily, sit relaxedly, or think quietly. Here the sun rose day after day on a mirror, or on gentle ripples, and the *Arcturus* pushed quietly and firmly through the ultramarine, fretted here and there with the ripple chains forged by flyingfish tails, or the great splashed stars where a tunny or dolphin leaped. Our night hauls were rich, full of new and exciting treasures, taxing our utmost time and energy to watch, describe and preserve.

[1] Galápagos: World's End, p. 163.

Early on the morning of the third day we were
up ready for the Humboldt Current, for new deep
sea fish, for wonderful floating things—for anything except what actually came to us.   At seven-
thirty, after sounding, temperatures, and breakfast, I went on the bridge and saw a very distinct
line in the water to the north.   The captain said
we had been steaming parallel to it since dawn.
I had the *Arcturus* turned toward it at once, and
found the Sargasso Sea of the Pacific, only in this
instance it was a wall of water, against which all
the floating jetsam for miles and miles was drifted
and held.   There came into my mind at once the
Humboldt Current, but I soon found that, most
astonishingly, that Antarctic river had nothing at
all to do with this gigantic Current Rip, which was
caused by the coming together of two warm, west-
wardly flowing streams of water.   When we first
detected the rip we were in 2° 36' North Latitude, and 85° West Longitude, which placed us
about two hundred miles southeast of Cocos Island.

When I approached within the possibility of
more accurate examination, I saw that the line,
which stretched from horizon to horizon, extended
in a northeast and southwest direction.   On our
side, the south, the water showed dark and rough,
but much lighter and smoother to the north. When
the *Arcturus* was at last actually astraddle of
the rip, I saw it as a narrow line of foam, zigzagging across the placid sea, with spouting white-caps
shooting up through the froth that marked the
meeting place of the great ocean currents.

The birds were the most noticeable inhabitants of this world of two dimensions, boobies of several species, stormy petrels, tropic and frigatebirds, soaring or feeding. Still more interesting than these was a flock of about two hundred northern phalaropes, strange little sandpipers which nest in Alaska and spend the entire winter far out of sight of land. These were massed in a close flock and flushed time after time just ahead of the steamer in the line of the rip. When finally they went on ahead for a half mile, they followed exactly every zigzag of the line of foam, keeping precisely to each bend of the denticulation of the current juncture. Twice after this I saw several of the little chaps cheating us of our belief that they never touched land except in the far north to breed, for they were perched on floating logs, picking out edibles from the crevices.

During the last few days we had observed a fair number of sea creatures, but here was a concentration of organisms greater than I have ever seen—the larger dotting the water and making visible its depths, the minute so abundant that in places they were of the consistency of soup. We had to give up trawling with the silk nets for two reasons; in the first place the throw and shift of the currents was so strong hereabouts that the nets and lines were often swept beneath the keel and in dangerous proximity to the propellers. Again, the amount of floating organisms was so great that the silk bags would fill immediately with a weight which strained them to the utmost. A few scoops with a hand net would collect a mass equal to a long haul through average ocean water.

When I realized to the full the significance of this tremendous phenomenon, I determined to spend a day or two in following the current rip slowly along, studying it as I went. Within a half hour of our reaching it a mighty school of dolphins came down the line, five or six hundred of them, leaping and playing, jumping high into the air, and presumably feeding as they went. For a while their long-drawn-out front, with its continual spouts of spray thrown high in air, looked like a counter current rip, extending in another direction.

For the first time I fully appreciated the advantages of the many strange contrivances I had invented for reaching down or getting close to the water. The pulpit now came in for constant use. In the Atlantic we had usually to keep this affair high above the surface, for the *Arcturus* would plunge and dip her nose so deeply that unless it was swung well up, one ran the danger of being washed out of it. Here the comfortably roomy iron flooring with its waist-high railing extending all around it, was lowered until it was almost at the surface, and here with harpoon or dip-net one stood, approximating the wonderful experience of St. Peter, at least in the early stage of his experiment.

From the stern of the vessel the crew had a veritable portière of hand lines baited for fish of all sizes from triggers to sharks. The gangway was lowered until the bottom step was awash, while on the port side, the boom-walk was perhaps of all the most popular and valuable point of vantage. Here we could walk easily along the double duck boards, with a guardian boom on each side, to a distance of thirty feet beyond the side of the ship, and lie down or sit or stand, with as excellent a view of all that went on in the water beneath as could be imagined.

I was astonished even before we reached the rip, to see logs of wood passing, many of them covered with an ivory mosaic of barnacles. Our pent-up energy had to find a vent in some way, and when I called out for volunteers to help haul one of the logs from the water up to the boom-walk, the instantaneous response together with the violence of the several attempts, warned me that this was the time and place where the static energy of my crowd was about to become transformed into muscular action. There is no precedent to be followed in the matter of getting floating logs on to boom-walks and so to the deck, and doing so without losing the inhabitants of the log. In fact, there had never been a boom-walk before, so it was anybody's method, failure or solution. Six of us began enthusiastically to collect the first log in the world ever thus to be gathered. As instrument after instrument proved inadequate, more material was shouted for and over the rail there poured a barrage of wire loops, boat-hooks, gaffs, nets and bags. One of the most enthusiastic of the loggers dropped two poles, a gaff, a bag and a net overboard and then went over himself to salvage what he could.   Meanwhile we had roped and wired the great mass, and by hanging by our knees and heaving willingly but all at different times, we got it up at last, dripping water, fish and crabs, and with a final shove heaved it over the rail to the deck.

I was afraid that all of the small people in the wooden sanctuary must have fallen out from the shaking and the banging to which the log had been subjected, but little did I know the clinging powers of these small beings. In the case of this particular log they might all have come of the race of Jumblies, for boring worms had been at work on it, perhaps when it was a pile of some far distant wharf, and by their activities had made half of it a veritable sieve. The long list of passengers would be out of place here; suffice it to say that we got fifty-four species from this single log. No sooner had we dumped it on the deck, than those of its inhabitants who objected most to fresh air began dropping off, first a five-inch trigger fish, followed by some younger brothers, and later a swarm of little blennies to whom the log must have meant much. For these fish are on their way to become quadrupeds of sorts, and are ordinarily never found far from solid shore. These belonged on the coast of Mexico, ranging as far south as Panama, which gave us at once a clue as to the origin of the current flotsam. They skipped alertly about on the deck, going where they wished, not, as with most fish out of water, where their flops took them.

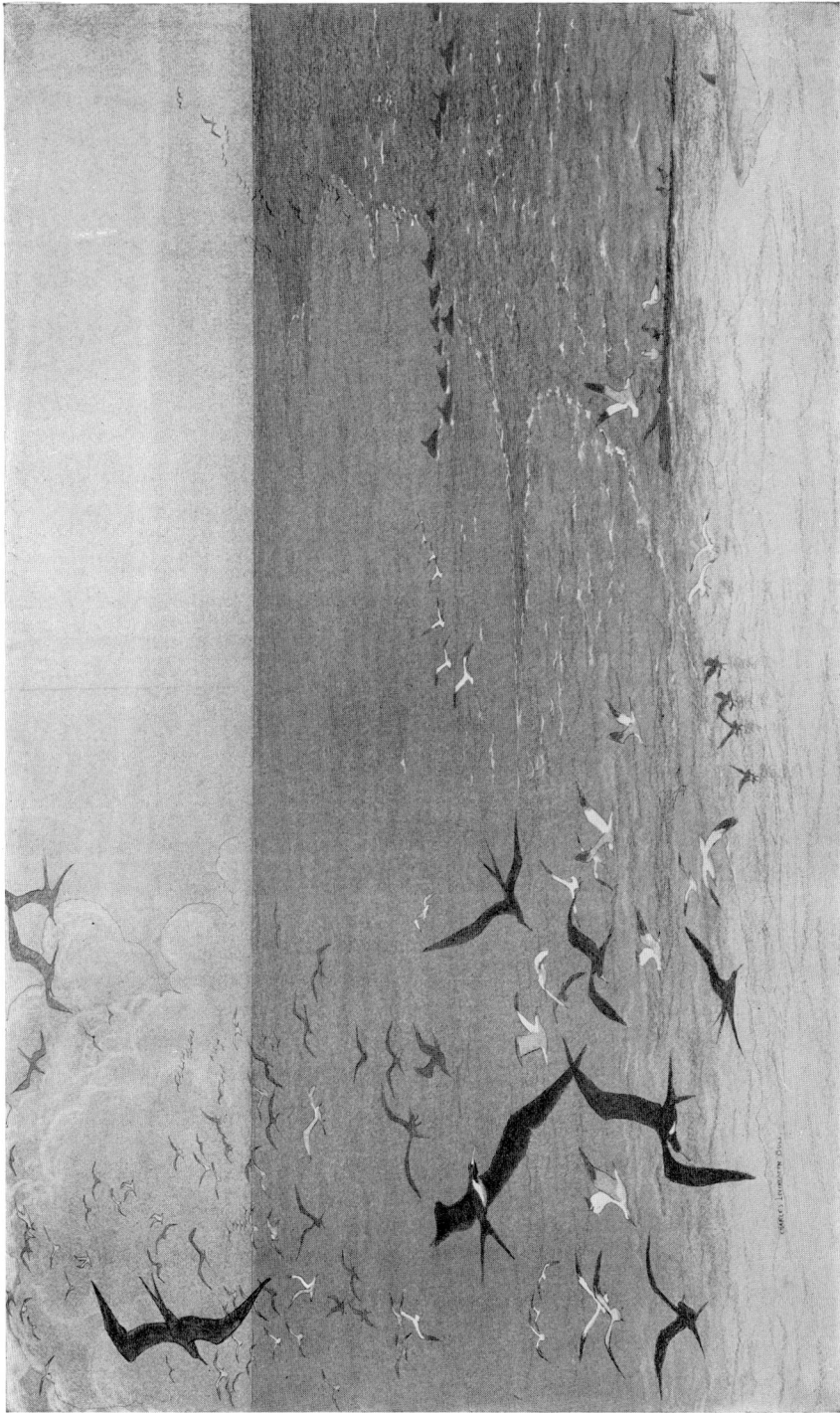

Fig. 12.—Life of the Current Rip.

Frigatebirds, Boobies, Gulls, Petrels, Dolphins, Sharks and other fish.

FIG. 13.—CURRENT RIP.

Encountered three days out from Panama.

At first glance they appeared black, but on close examination showed a glory of scarlet spots all over the head and pectorals, and maroon and sage broken bands on the body, with the median fins varigated yellow and red. Over the eyes were two long, lemon filaments, and a blood-colored Y-filament at the nostrils. They looked intelligently about with their pop eyes, and lived through vicissitudes which destroyed all other fish.

Crabs in multitudes crept about or were picked out of crevices and water-worn cracks. Some were pale olive-gray, irregularly mottled with maroon, looking like bright-colored conglomerate rocks. On the legs were sea-green swimming fringes. The ivory-white under parts never showed, as the crabs always scurried about with bodies held close to their pelagic island. Some of the forward-bent abdomens were cupped about a large mass of chocolate spawn. Other species of crabs were deep, Dutch porcelain blue, and one dark chocolate one had a big transverse rectangle of white like the sargassum crabs.

The log reminded me of a large piece of fossil rock, such as I used painstakingly to hammer out of New Jersey quarries. Wherever a knot had rotted away, or a teredo worm had gnawed out a tunnel, the interstice or crevice was filled up by an animal which fitted as if it had been poured in, a kind of living fossil embedded in the dead wood. Especially was this true of enormous worms countersunk in every possible crack. These were seven and eight inches in length, with nu-

merous bunches of curly medusa heads of reddish
tentacles above, and dozens of brush-like tufts of
white spines.

Again and again I was impressed with one outstanding feature of the Current Rip, this uncharted zoölogists' paradise—the narrowness of its
limits and the sharpness with which these limits
were defined. It was a world, not of two, but to
all intents and purposes, of a single plane—length.
From first to last we followed its course along a
hundred miles, and yet ten yards on either side of
the central line of foam, the water was almost
barren of life. The thread-like artery of the currents' juncture seethed with organisms—literally
billions of living creatures, clinging to its erratic
angles as though magnetized. The floating, drifting world of ocean life was, of course, irresistibly
swept there, and this life alone would have made
it worth a year's study. There is no stronger attraction in life, however, than food, and here was
food, manna, ambrosia, in stupendous quantity, to
be had for the taking. Somehow the news must
have spread far and wide, over and through the
great lake-like expanse of this part of the Pacific.
As each group and individual sensed the happening,
another and still another one or one thousand, a
little farther away, saw the eager start and in turn
started. I can in no other way account for the
infinite number of fish and organisms other than the
helplessly drifted plankton which filled all this rip.
It seemed as if as great an area must have been
depleted of its larger, self-swimming, dominant

creatures, as of the lesser, wind-driven, current-swept folk.

These last helpless ones have been given the name plankton, which is appropriate, for when the Greeks used it, they meant Wanderer. Here we saw what must have amounted to many, many tons of these minute beings—diminutive crustaceans, both adult and larval, the myriad species of jellyfish and pelagic mollusks, worms, larval fishes, single-celled animals such as those which light up the sea at night, and my jolly little friends, the flying snails. Where these are gathered together in numbers, there will the self-determined fish be, tiny little chaps who dash about and feed upon the living soup of the sea. These in turn, attract middle-sized fish, and these still larger ones. This would seem like a straight line—a linear chain of life, but it is, in reality, a great segment of a curve, the circle being completed when one of the great marauders dies, and furnishes food, not only for his former victims, but for the minute creatures that he would have disdained as nourishment.

Although compressed within so narrow a longitudinal area, yet the slow procession of the wonderful fauna was far from uniform. Whether we use the simile of corpuscles tumbling along a stream of blood, or some less apt memory, the nodes in the line of life were the logs and other débris. The number and diversity of these were beyond belief, and I longed for a botanist to identify them all and perhaps to tell from what exact coastal or river forest or jungle they had drifted. Of one

thing I was certain—all were tropical. None had
come from more temperate regions, borne along on
that Humboldt Current of which as yet we had
found no trace. I remembered the sentence I had
written in my Galápagos book, sponsoring the continental origin of that Archipelago:

"As with my theory of the origin of flight through *Te-
trapteryx* and my classification of Phasianidae by tail
moult, so with all my points of view which in our present
state of knowledge must be wholly or in part theories,
I hold them in readiness to be relinquished at the first
hint of better proof on the opposite side,"

and wondered whether this Current Rip must be the
opening wedge to relinquishment. It was powerful evidence for the opposition—those who held
that the Galápagos had always been isolated
islands, planted and populated by the accidents
of drifting seeds and transported insects, birds and
reptiles. Here I was, just about half-way between
the outermost headland of Panama, and the out-
lying island of the Galápagos, and, passing slowly
but steadily to the southwest, was floating jetsam of a size sufficient to support any member of
the Galápagos fauna, jetsam laden also with seeds
and sprouted plants enough to suit an island-
favoring botanist. Within an hour, there passed
log after log, sticks and solid pieces of wood, besides three bits of wreckage from ships. I noticed
a forty-foot Cecropia, six inches through, bamboos
up to five inches diameter, and soft, pine-like wood,

besides sections of palm trunks and a cocoanut in the husk—all rotten, all alive with living creatures catching a ride. During my stay, I made a list of thirty-eight species of trees, plants and seeds, and of thirty-two of whose identification I could be reasonably certain, not a single one is to be found in an exhaustive list of the flora of the Galápagos. Either this marvellous Current Rip is a recent phenomenon, dependent in some way upon the inexplicable shifting or absence of the true Humboldt Current, or its course, beyond where I could see it, was deflected. Both, indeed, may have been true, but of the former I have no means of judging. To anticipate our movements, I may state that after remaining and studying the rip for two full days and nights, I followed it for several score miles, and, as I shall narrate, saw it turn steadily northward, until, at 2° 8′ North Latitude, and 86° 4′ West Longitude it was headed west by north, by one-quarter north. If it only maintained this direction it would clear the northernmost island of the main group of the Galápagos by one hundred and fifty miles, and even the most northern of all, the isolated speck of Culpepper, would be a full hundred miles south of the influence of this log-rolling current wall. So, at least from this angle, my theory is still perfectly tenable.

Four large sharks loitered around the ship in most deliberate fashion, and there was a wild scurry for harpoons. John Tee-Van, descending to the pulpit, brandished one of the weapons to

an accompaniment of jeers from his observers.
They discovered, however, that it is not safe to
predict failure merely from the premise that the
venturer is an amateur. With as much precision
as though he had made a life-long study of harpooning, he hurled the spear not only into, but
straight through the shark and the half-hour
struggle to hold the creature was sufficiently exciting to satisfy the most exigent of big-game fishermen.

The other three sharks were not alarmed by the
fate of the first. They lingered on the scene of
his disaster, and from the boom we paid out string
with pieces of meat for bait. They came as easily
to this toll as a donkey following a proffered car-
rot and by pulling in the tempting morsel two feet
in front of the eager blunt snouts, we brought them
to the surface directly under our feet, so that we
could watch the movements of the brilliant blue
pilotfish, that, with uncanny prescience, anticipated every movement of their huge patrons. One
of the big fellows had three of these little satellites
that unfailingly held their formation, one just
above his head, the other two in perfect alignment
a few inches in front of his jaws. So exactly
synchronized are the movements of such a marine
galaxy, that it is impossible to tell whether the
shark follows the pilotfish, or the pilotfish the
shark. It is evidently a profitable arrangement for
the pilots, since we meet with few cases of philan-
thropy in marine life, and whether they actually
lead the sharks to food, or are merely hopeful

hangers-on, at any rate they must benefit by the crumbs that fall from the sea-wolves' table.

The sharks had even more literal hangers-on, in the persons of the shark suckers. The big fish can seldom be lonely, for there is scarcely a shark to be found without at least one of these pseudo-parasitic attendants, known as *Remora* or *Echeneis*. Clinging with the great sucker which has, in some way, evolved from the dorsal fin, these strange creatures can slip at will over the whole of the shark's body. When their host is hooked, they cling until the very moment when he is drawn into the air. Then, realizing that the worst has definitely happened, with an admirable expediency they desert, not the sinking, but the rising ship, and hurry away to find some less unlucky means of transport, whether shark, or, it may be, some other great fish or a turtle. We took two *Remora* with hook and line, which is rather unusual.

Late in the afternoon of our first day in the rip when we had stopped in order to take temperatures, I was looking down from the bridge when I suddenly saw a sea snake swimming in small circles and drifting slowly along. It recalled the last meeting I had with these real sea serpents —when I balanced in the bow of a sampan in the swift running tide of Penang. A Chinaman steadied the boat for me with his long sweeps, while I dipped up various desirable creatures as they swept past on the current. As I had no bottle or bag of sufficient size I carefully avoided the sea snakes which were swimming past, literally in hundreds. They were brilliant in color, olive green above, with many, broad, yellow cross bands, about as protectively colored as yellow daisy blossoms in a green field.

I knew they were also found in the eastern Pacific but had not seen them here before, and I keenly wished to capture this one. Two of our small boats which were overside, were too far away to understand our frantic signals, so, handicapped by the thermometer line being out, all we could do was to hope that the reptile would drift down on the ship. Luck was with us, for while we watched breathlessly, our first sea snake writhed so close under the boom-walk that we were able to scoop it up with a long-handled net. Before the net closed over it, it seemed to be biting at a part of the body where I could see a small white spot.

I seized it back of the head and dropped it into an aquarium, taking considerable care in the process as these are as poisonous as any of the venomous terrestrial species. It did not struggle much or seem to have the strength which a snake of its size—almost three feet long—should have. From the water of the tank it lifted only the head and neck, and showed no interest in its new environment. This lethargy was doubtless due to two severe bites which it had received from some foe. At one of these it had itself been striking, probably in unreasoning irritation at the pain. It had several patches of good-sized barnacles along the body, and some small ones even on the crown and chin. Nothing with a scaly or a hard skin seems safe from these omnipresent crustaceans. I once thought that after they had grown for a time, they must set up a certain amount of irritation, but I have removed barnacles of good size from fish, without finding any trace of lesions. Here too, when I scraped a few off, neither in surface or pigment was there any alteration noticeable from the normal.

I had this *Hydrus* painted, photographed, and his method of swimming studied, then chloroformed him to put him out of his misery. He had been feeding on two young Coryphænas—the dolphin-fish of the ancients, which we found so abundant hereabouts.

This individual was quite as brilliant as my Malay species, but absolutely unlike it in pattern. The dorsal third was black, and the ventral surface and much of the lower sides olive-green. Between the two colors ran a broad band of bright chrome yellow. On the long, flattened tail, this latter tint dominated as a background, over which were scattered a number of large spots and imperfect bands of black.

Besides the sharp keel to which the body narrowed below, and the paddle-like tail, these snakes are so intimately associated with an aquatic life that they cannot survive protracted removal from it. Why this is, no one has had sufficient curiosity to ascertain. Its breeding habits are said to be like those of the seal, as it is viviparous, and goes ashore to bring forth its young in the crevice of some great boulder. A large female was once found in such a place, coiled about a score of young, each of which was two feet in length. We caught two more snakes in the Current Rip, and saw a number of others which dived at our approach. Without exception all we caught or saw were parasitized by the barnacles, one having twenty-seven clumps. These were all of one species, stalked, the shell being a delicate maroon with two Y-shaped white markings (*Conchoderma virgatum*). When I had several snakes for comparison I saw that the tail pattern is not only wholly individual, differing in each snake, but the pattern varies on the two sides of the tail in the same individual reptile.

I started a trawl with several metre nets at various depths, and leaving directions for the *Arcturus* to revolve in a five-mile circle, I went overboard with John Tee-Van in a small boat and for several hours we rowed about in this astonishing longitudinal maelstrom. I cannot recall having ever seen so many living creatures in so limited an area in all my life. In the distance dolphins still splashed and sighed, boobies whistled by and dived like plummets, gulls and frigatebirds picked up bits of their choice with graceful delicacy, now and then a turtle drifted past, or dived and watched us from beneath our keel.

Sharks occasionally swam by, and twice, by intention or accident, one bumped into our skiff. Later in the afternoon when Dr. Gregory was out, a big shark followed his boat persistently, circling often, and repeatedly bumped so hard against the

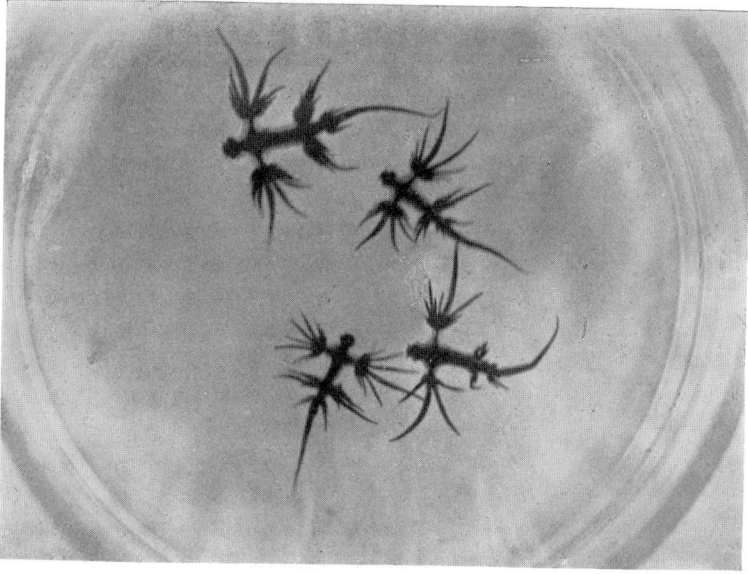

FIG. 14.—GLAUCUS.

An ultramarine sea-snail without a shell, living at the surface.

FIG. 15.—PAPER NAUTILUS.

A cousin of the Octopus which lives in a delicate, tissue paper shell.

FIG. 16.—EGGS OF THE INSECT *Halobates*.

The Water Strider of mid-ocean lays its eggs on the floating feathers which have fallen from the wings of Gulls and Boobies.

FIG. 17.—LIFE OF THE CURRENT RIP.

Thousands of living creatures taken with a single scoop of this tub. Most distinct are Jelly-fish and Floating Snails—*Porpita, Glaucus* and *Ianthina*.

boat that they were rocked and jarred, and not having even a boat-hook, they began to row back toward the ship. I never heard of such a happening before.

What looked like oval, thick, greenish cigars were floating pelagic anemones, mouth down. At the top a small group of white bubbles—the float— then a circular, dark-green, caterpillar-like body mass, below this a ring of numerous, short, white tentacles, and finally, at the bottom, the expanse of greyish tissue about the mouth. They looked like strange swollen green acorns, with a white stem base and white cup.

Although I have said it before, I must reiterate that the teeming amount of life was unbelievable. Two dips with a butterfly net yielded half a pail of organisms. In one place the water for ten square yards was tinged with deep purple, thousands upon thousands of tiny salpas, each with its large nucleus. The most consistently abundant things wherever we rowed were uncounted myriads of small, rounded, pale spheres, which proved to be the eggs of some unknown species of mollusk.

The strictly surface life was as teeming as that beneath. In the bubbles and spray strung out along the rip were hosts of oblong patches of finer froth, and suspended from one end of this, was always a beautiful purple-shelled *Ianthina* snail. Almost as numerous, and often in solid masses, hundreds of the strange tufted nudibranch, *Glaucus*—dark ultramarine above, shading into mother-of-pearl on the arms, and to ivory white below, looked like an azure-fringed frog, or some distorted fleurs-de-lis armorial bearing.

Porpita was abundant—those little floating colonies of animals, which I have seen even off the New England coast. At a distance they look like either quarters or silver dollars, according to age, but when I sit down in front of one floating in a glass dish, descriptions and similes pall. On my laboratory table is a beauty with a disk two inches across. I have seen unbelievably minute crystals of some rare mineral, or a thousand beams of sunlight radiating over still water which reminded me of this, but the delicacy of color and pattern are beyond all verbal or written appreciation. The center is yellowish gold, and from here to the periphery, about one hundred and fifty lines radiate and undulate. It is crenulated and waved, and the pale blue and dull yellow are inextricably mingled. The broad margin is deep, deep blue, and outside there are three to five ranks of delicate tentacles. Their long stems are beryl blue, while the rounded beads which double-line the tips are of the darkest ultramarine. Such is a hint of the beauty of one mote among the trillions on every side.

Near the side of the skiff I saw a small white creature dart away, spread four wings with a black spot in the center of the hinder pair, rise and fly for a yard, then drop, and again make a short flight. It was so like a butterfly that for a moment I was too astonished to move. Then I called out, pointed to the tiny flyingfish and my companion caught it. If it had kept quiet we should never have seen it among the spots of foam. Putting one's hand down into the water was to feel a host of creatures, some visible, others not to be seen until they crashed on the vision in a dazzle of iridescence.

In some old magazine of natural history there is a report of the eggs of *Halobates,* the water striders which live on every ocean, being found on a floating feather, but, as far as I know, there has never been a reconfirmation of this. In the course of our association with the Current Rip we found, not one, but seven examples of it. As we were rowing slowly about, I saw a long white wing feather of a booby, which seemed to have some strange encrustation. I scooped it up and found that three-fourths of the vane was clotted with a rust-colored mass of ova. I did not stop to examine this carefully at the time, as new specimens were passing at every moment, but put it in a small aquarium of running water. The next morning both this aquarium and the four succeeding ones were a maze of tiny skating figures, and the distended stomachs of the small fish in two of the tanks, showed that others than myself appreciated this discovery of hatching *Halobates.* I found that there were at least twenty thousand eggs on one feather, undoubtedly representing the united efforts of many females. Some of the eggs seemed newly laid and these would often overlap others that held large embryos. Under the lens they looked like a mass of tiny grains of rice, some tan, some orange in color. Two more feathers were taken later, and four large ones were seen passing, all heavily laden with the hemipteran ova. Outside the rip I noticed four additional lots, in the course of this trip, three on feathers and one on a piece of wood. Nine out of the ten feathers were white ones from the wings of boobies, the tenth was brown, probably from an immature bird of the same group.

From the small boat on the same day we were fortunate enough to catch in a pail one of the enormous, smoky-grey egg masses, a dozen of which I had seen floating by the ship. In a glass aquarium it looked like some loose-textured sponge, with great openings here and there like the vacuoles in a sponge. The microscope showed vast numbers of small fish eggs—a small bit teased into a watch glass contained twelve hundred and seventy-six. I was greatly disappointed at not being able to rear some of these, but the aquarium pump went wrong at this time and these, among other specimens, were destroyed. Our curatrix of larval fish had better luck with a few in a dish and kept some alive for seven days. Certain characters seemed to stamp them as young *Coryphœna,* but we could never be quite certain.

The dominating fish of the whole Current Rip were unquestionably young amber-jacks or yellow-tails, the well-known game fish of the Pacific coast. These were present in schools of tens of thousands, each school keeping in dense formation, and moving with that inexplicable unanimity which has made me so often use the expression, the spirit of the flock or colony, herd or school. There would sometimes be several hundred of these fish massed under the keel of our little boat as we rowed about. They refused all bait and it was with great difficulty that we secured two or three specimens.

We had been less than a week out from New York when we discovered the value of the gangway as an adjunct to night fishing, and although we had made use of this on all occasions, we had no hint of its real possibilities until now. At dusk, when the *Arcturus* was safe cradled between the two pressing walls of water, I had two clusters of electric lights lowered to the last steps of the gangway and focussed down upon a twenty-foot circle of water. To sit and watch the gradual concentration of the ocean life attracted by the light, was to have a very wonderful experience.

The first arrivals were *Halobates*—the water-striders of the sea. Two years before I had found their newly hatched young in thousands close to the shore of Indefatigable,[1] and today I had verified the secret of their cradles. A hundred soon gathered and covered the surface of the lighted area with a maze of shooting lines.

No amber-jack came, but *Coryphæna* was there in numbers, and we caught thirty or forty, all less than a foot in length, reflecting every imaginable color. This marked the beginning of the inevitable chain of reactions—first the small fry and then the small fish; next the outposts of the mighty army

[1] Galápagos: World's End, pp. 83-86.

of the middle-sized—the mid-links, feeding upon and fed on. After the *Coryphæna* and others of their kind had played about for a while, faint, ghostly shapes began to appear far, far down, and soon a shark rose to the surface, and nosed about to see what this new thing might have in store in the way of crippled or dead. The most exciting visitors—and they came in all sizes and colors—were the squids. In and out weaved little two-inch chaps, pursuing fish of equal size with such speed and ferocity, that when one leaped, they both leaped out of water. My net would slip under a scarlet squid, and in the length of time it took to lift it to my eyes, the net appeared empty, until a slight sag in the mesh showed where there lay a squid of pearly whiteness.

A six-inch species placed in the big tank gave a most marvellous exhibition. From side to side it darted so swiftly that the eye could scarcely follow, and at the end of each dart, as it brought up against the glass side, it was a different squid— first scarlet, then salmon, rose, scarlet again, pink and the white of a moonstone.

We had to have a clearing house, or rather a clouding house aquarium for newly caught squid, in which, as soon as deposited, they could empty their sepia bags. A big squid three feet long which we harpooned, ejected an enormous quantity, not sepia, but opaque, bluish brown ink, that gave off reddish bronze reflections like the skin of his body.

If in the permanent aquarium with the larger squid, there remained by chance a hapless fish, a moment after the first frenzy passed, the squid went for it like a flash of lightning, seized it, and hugging it close to the heart of the horrid circle of arms, began to devour it, always beginning at the throat.

A passing swell, coming out of the black night, would fill the lighted circle with a mêlée of jetsam—porpita, ocyropsis, ianthinas and salpa, which, if you do not know them by these names does not matter, for if you will allow your imagination full play, and try to think to what strange and beautiful beings such names might apply, your mental images will yet fall short of the strangeness and beauty of the reality.

At a critical moment of the fishing, when we were keyed up for something great and weird, there flew into the glare a fluttering school of the little, snowy-winged, butterfly flyingfish. A villainous atom of a blood-red squid shot forward at them and three flew straight into my net. Large pelagic crabs came and went, wine-colored, with purple swimming legs, eyes wavering on long stalks, and long, many-toothed claws, waiting for what the squids did not get. Half-beaks shot across the circle, as rapidly as the squids, and half-transparent fishlets showed first one, then another outline as the light and waves partly revealed them. The greatest surprise was when a very large silver hatchet fish, *Argyropelecus,* floated into view. It was dying, as it had been badly bitten by some creature, but it was the first and only time I ever saw this richly luminous fish at the surface of the sea. Not many miles away I was later to take one in a tow net at three hundred fathoms, but the center of their distribution seems to be in still colder, darker water, about five hundred fathoms, two-thirds of a mile down.

Nature loves contrasts, and close on the passing of the flock of white-winged flyingfish, a great creamy white shape appeared and vanished again far down in the translucent depths. Then it rose head first, a large shark as we thought, heading straight for the gangway. Just before it broke water, someone shrieked "It's a squid!" and at the word half the monster shot into the air, his wriggling tentacles seeming to reach for the row of legs that dangled from the ladder. A chorus of excited shouts arose from the four of us who were on the spot, an inadequate harpoon splashed harmlessly beside him, and the creature dashed backward and sank out of sight. He was different from the other lesser squids, not only in size and shape, but in color, being a pale pinkish tan wholly unlike what any of the others could achieve by whatsoever combination of their chromatophores. Hardly had we gasped out our joy, when in exactly the same spot he appeared again, and went through the same manœuvers, springing from the water as though propelled by a submarine cannon. Allowing for every illusion of night, water, light and excitement, the most conservative estimate placed his length at eight feet, and the width of his body at nearly two. None of us will ever forget the spectacle of that long, torpedo body shooting out of the froth of the rip, the snaky, outreaching arms beaded with big vacuum cups, and above all, the huge disks of eyes which glowed like silver plates in the tan flesh.

The afternoon of our last day, the life of the rip seemed, if anything, to have increased. Full grown *Coryphæna* played about, and now and then we hooked one, but they were usually too strong and heavy to be played successfully from the boom or the deck. Seen just beneath the water, they were a blaze of color—the body emerald, the pectorals turquoise and the tail clear yellow gold. Sea snakes undulated past, their golden spotted tails flashing out as they turned and looked up at us. Great turtles drifted along, as motionless and as barnacled as the logs about them. A dolphin-fish leaped over one and darted about it, but the turtle looked only at us. Another *Coryphæna* dashed by with a great piece bitten out of its shoulder. I cannot imagine what enemies these high-powered engines of the sea can have, except real dolphins, unless they wage battle with one another.

From the deck, looking directly down, we could watch clearly the fish which crowded beneath every log, or stick or nut. Big triggerfish, over a foot in length, often lie flat on the logs, half out of the water, or jam themselves into crevices in attitudes most astounding for a fish. Once a twelve-foot hammerheaded shark swam slowly around the whole ship.

We dared not go below for a moment, and begrudged every minute at meals, for fear we should

miss some of the absorbing tableaus and exciting
events which were constantly passing on every side.
Always the key to the meaning of the actions and
reactions lay in the complex inter-relationships of
all these myriads of living beings. We were trawling slowly ahead, and bore down on a small log on
which was perched a booby, a big fellow in brown,
with pink bill and greenish feet and legs. Just
before our bow upset his raft, he ejected three fish,
each at least eight inches long, and so recently
swallowed that they seemed still living. Immediately every fish under the log dashed at this sudden
appearance of food, and for yards around the excitement spread and spread. Meanwhile the
booby, lightened by the discarded ballast, flew off,
spattering the water with his great webs for a few
yards, and followed eagerly a little way by the
hammerhead and several dolphin-fish.

Just after this we caught a *Coryphæna* from the
boom-walk which weighed thirty-two pounds and
as completely disintegrated the white sunlight to
our eyes as any prism which ever reflected a spectrum. From its gills I took twenty-five parasitic
crustaceans, of which half were in turn parasitized
by pinkish goose barnacles. And this great fish
had been feeding on the most beautiful sea-shell
in the world—a dozen paper nautilus, all uncrushed
and unharmed, with their little argonaut owners.

Three times on this last day great areas of the
water were colored a deep purple by incredible
numbers of delicate jellyfish, and again a yellow
stain was spread over a hundred yards of surface,

BUTTERFLYING FISH

Cypselurus furcatus (MITCHILL)

Young specimen
(Twice natural size)

PLATE II

—billions and trillions of microscopic creatures manifest to us only by a tint. I fished up a many-branched bamboo tree-top about twelve feet long, which, for many seasons, must have waved in the breeze fifty feet in the air in some distant jungle. Now its slender side branches all seemed in full flower, tipped and beaded with a myriad ivory barnacle cups, swaying on their little stems, the whole looking for all the world like a gigantic spray from a Japanese cherry-tree in April.

As to the physical reactions of my great Current Rip, at five-thirty on the afternoon of the first day we steamed into the center, faced eastward, which was up wind and up current, and there lay all night. During that time we drifted eleven miles to the west, the current being about one and three-tenths knots, and we turned completely around twice, but never left the heart of the rip. We rolled slightly all night and three times I was awakened by what sounded like breakers, which proved to be the rip near by, the sea in the distance showing calm and quiet in the moonlight.

During two days we repeated this experience three times, with the invariable result of swinging up wind and current, then vibrating slightly from side to side as first one, then the other current pushed us toward the dead center. I tested the temperature a quarter of a mile on each side of the central line, and found that the southern current was four degrees colder at the surface, and two degrees as far down as five hundred metres. The current on the south side flowed about two and a half knots, and that on the northern side one and one-half knots.

The orientation of our mean position in the rip was,

> South from Cocos—210 miles
> Southwest from Panama—400 miles
> West from Ecuador—340 miles
> Northeast from Tower Island—300 miles

The trend of the rip on April first in 2° 5′ North Latitude, and 85° 53′ West Longitude, was southwest by west one-half west (242), and by the afternoon of April second in 2° 8′ North Latitude, and 86° 4′ West Longitude, had swung to west by north one-half north (285), a shift of forty-two degrees to the northward.

The last view I had of the rip was a dramatic and memorable one; five great, red-footed boobies perched close together on a floating log. Four were in the white adult plumage, and the fifth still in immature brown. They paid no attention to us, although they were less than fifty feet away, doubtless considering the *Arcturus* merely a log of greater size, and us, marooned fellow birds. As I watched, the fin of a gigantic shark circled close to the log, passing completely around it four times. The birds paid no more attention to this than to us. Slowly they drifted past and the sunset stained their feathers a delicate salmon; then night and distance swallowed them up, and I shall never know more, either of the satisfying of the appetite of the shark, nor the slumber and dreams of the sea birds.

CHAPTER III

WITH HELMET AND HOSE

I AM twenty feet under water with a huge copper helmet on my head, tilting with my trident against an olive-green grouper over a yard long, who is much too fearless and inquisitive for my liking. Not until I have pricked him sharply with the grains does he leave off nosing my legs with his mean jaws and efficient teeth. It suddenly occurs to me how knightlike I am as far as the metal casque goes, and then in spite of the strange world all about, my mind goes back to the long-ago Christmases when a new-published Henty book was an invariable and almost the best gift. I instantly know that if ever I succeed in shackling these divings to mere, awkward words it must be called "With Helmet and Hose," and if any modern boy, grown-up or gentle reader does not know why, explanations will do no good.

I wish I could credit my present passionate enthusiasm for diving beneath strange tropical waters to a life-long suppressed desire—an *idée fixe* which would not be gainsaid. But unfortunately this is not so. My only excuse is that I suffer intermit-

tently from what my artist once offered as a definition of a monkey, a desire to be somewhere else than
I am.

Considering carefully this whirling ball of mud
upon which I found myself, I read in books and
saw pictures of jungles and deserts, and my desire
to see them was just a little stronger than the many
obstacles between; I had breathed the air and
watched birds fly for an unconscionable number of
years before I began my first wobbly taxi-ing
across a flying field. Since then I have left the earth
under pleasant and unpleasant conditions over three
hundred times, and, except twice, returned safely.

Without shame I confess that I have lain awake
nights and spent innumerable hours of my life in
gazing at the moon and planets—nay, even at the
Small Magellanic Cloud with desire and longing,
for if one wishes to visit inter-stellar space, one
might as well hold the thought of a passage on
Tomlinson's route as on a mensurable moon trip.
Up to the present, twenty-two thousand feet is as
far as I have been able to rise above solid ground.

Another realm which has always seemed as re-
mote as the moon is the depth of the ocean. My
reading and wishing never took any concrete, definite direction until the trip I made to the Galápagos on the *Noma*. Then I first realized the glories
and desirability of the submarine world. This at
once encouraged and then disheartened me—the
encouragement coming from the ease of diving from
a boat or a pier and watching for a brief moment
the fish and sea-things, simultaneously with the

realization of the futility of such a brief, blurred glimpse.

I inspected a number of diver's outfits one day and found nothing tempting in the enormously cumbersome suits. Then, just before I sailed on the *Arcturus,* I bought my helmet. The paraphernalia accompanying it were so simple that I doubted its efficiency, but at least it was an effort in the right direction of investigation of a new world.

During the first part of the *Arcturus* adventure the sea was too rough to think of using it, even a few feet below the gangway, but when we moored close under the cliffs of Darwin Bay at Tower Island—our old Galápagos anchorage—I brought up the box from the hold and unlimbered the diving apparatus. The helmet was a big, conical affair of copper, made to rest on the shoulders, with a hose connection on the right side and two oblique windows in front. Around the bottom extended a flange on which four flattened pieces of lead were hung, each weighing ten pounds. This made a total weight of sixty pounds for the entire thing. The hose, which was of the ordinary common or garden variety, was attached at one end to the helmet and at the other to a double-action automobile pump, which screwed to a board, and was operated by a long iron lever, pushed back and forth. Almost at once we elaborated a method of operation which was so simple and satisfactory, even to the slightest details, that no change was necessary after weeks and months of use.

Our regular mode of diving is as follows: We start out from the *Arcturus* in a flat-bottomed boat which has a square, eighteen-inch glass set in the bottom amidships. My regular diving crew is John Tee-Van and Ruth Rose and we three dived in many and in strange places. To the stern is fastened a long, metal Jacob's-ladder, rolled up when not in use. We are towed or we row to the shore, preferably to the base of cliffs or steep rocks, as that affords considerable depth close inshore and rocky places are beloved by hosts of fish. We anchor as close to the cliffs as is safe, and roll out the ladder, so that it sways in midwater or rests upon the bottom. The pump is in the bow, the handle fixed, and the leather washer carefully screwed in. The hose is cleared of kinks, and is looped, partly overboard. A hand line is tied to the top of the helmet, and the inside of the glass windows is coated with a film of glycerine to prevent the breath of the diver from condensing and so clouding it. The four lead weights are slipped over the flange on the helmet base and all is ready for the diver. A hand water-glass is near for constant lookout for danger, and one or two long-handled harpoons.

In bathing suit I climb down the ladder over the stern, and dip to my neck, being careful not to wet my head. Then John lifts the helmet; I give a last, quick look around, draw a deep breath, duck into it, and as it settles firmly on my shoulders, I climb slowly down. The sensation just above water is of unbearable weight, but the instant I immerse this goes and the weight of the helmet with all the lead is only a gentle pressure, sufficient to give perfect stability.   Meanwhile Ruth Rose has started the pump.

From a blurred view of the water surface and the boat's stern, I sink instantly to clear vision under water. I descend three rungs and reach up for the short harpoon or grains which is put into my hand. At the fourth or fifth rung the air presses perceptibly on my ears and I relieve it by swallowing. For the first moment there is a muffled rumble of bubbles escaping, which I never noticed until I heard Ruth exclaim about it. This ceases as soon as the helmet is entirely under water. I descend slowly, swallowing now and then, and when the last rung has been reached, I lower myself easily by one arm, and lightly rest on the bottom. If serious danger threatens or the pumping should go wrong for any reason, I have only to lift up the helmet, duck out from under it and swim to the surface. The level of the water keeps constantly at the level of my neck or throat, and if I lean far forward it gradually rises to my mouth. But there is no splashing, no sense of oppression.

In most of the great changes or experiences which come to us humans, such as seeing our first palm tree or circus or volcano, the first reading of Alice, diving, a battle, discovering the method of complete relaxation or really being *in* the only Borneo in the world, it is not, as so many people think, the first few minutes which are the most wonderful. It is the subsequent gradual appreciation which develops that realization of the wonder and the beauty of the thing close at hand. It is so easy to miss this almost conscious appraisement, and after the trip or performance or experience is past, we long for just one moment of the actuality, so that this or that could be seen again and remembered more clearly. Before I started on my trip around the world in my search for wild pheasants, someone gave me one of the most valuable hints I have ever had. It seems a foolish little game when I come to write it down, but it is based on a very sound realization of a great human weakness—the contempt bred by myopic familiarity, the absolute necessity for even an artificial perspective. It consists merely in shutting your eyes when you are in the midst of a great moment, or close to some marvel of time or space, and convincing yourself that you are at home again with the experience over and past; and what would you wish most to have examined or done if you could turn time and space back again. A hundred questions rush into this induced mental vacuum—what were the color and shape of the wild blossoms upon which that pheasant fed? What was the sound of the anti-aircraft shells? At what speed did the lava flow? etc.

And so, as I said, I swung myself lightly down from the ladder and stood on the bottom. I gazed out with interest on the rocks and fish about me, but felt a vague feeling of disappointment. I was breathing so easily; the water outside might have been correctly heated air as far as any bodily sensation went; I was looking through a pane of glass at fish swimming about—exactly what I have done and seen a hundred times in our aquarium in New York. I felt only as if I were in a very small, strange, but perfectly comfortable room, looking upon a wonderful tank of living fish with a most excellently painted background. The shock of entrance into this long-anticipated world had not been as radical as my imagination had pictured, even although I cannot recall having visualized instant attacks by huge sharks, or the feel of the snaky tentacle of an approaching great octopus. The fact of my bodily comfort and the vivid memory of aquariums all over the world had deadened the stupendous marvel of it all.

I sat down on a convenient rock, shut my eyes, and recited my lesson: *I am not at home, nor near any city or people; I am far out in the Pacific on a desert island, sitting on the bottom of the ocean; I am deep down under the water in a place where no human being has ever been before; it is one of the greatest moments of my whole life; thousands of people would pay large sums, would forego much for five minutes of this!*

This was enough. I opened my eyes and saw, resting on a rock not more than three inches away from my face, the red bull of Kim. It was the strangest little blenny in the world, five inches long and mostly all head, with tail enough only to steady him in his place on the boulder. His long snout with nostrils flaring at the tip, his broad, flat crown surmounted by two curving horns,

made him absurdly like a prize bull. He was dull
scarlet with splashes of golden brown along his
sides, which was well enough, but a bull does not
have tatters and fringes of blue and yellow scattered all over him (unless we choose to consider the
cruel banderillos as ornamental). My blenny's
eyes were silver with hieroglyphics of purple in
them, and as I looked, he puffed a puff of water
at my window and was gone.

I was quite reoriented now. The hardest thing
was to realize that I was *wet*. It was the old story
of the value of comparison. All of me was wet
and I could not reach up into dry air, so I had no
sensation of wetness. I looked at my fingers, however, and saw the beginning of washerwoman's
wrinkles, so was convinced! I reached out and
picked a starfish from the rock in front, and as it
slowly crawled over my hand, I realized to the full
that this was a wild starfish and not one brought
from somewhere else and placed there for me to
look at.

One handicap, present at every submersion, was
the impossibility of writing down notes, except on
an awkward slate, the multitude of exciting experiences and hosts of remarkable creatures so dis-
tracting my attention that my memory was strained
to the utmost to recall a clear sequence of events.
This I hope to remedy in a made-to-order helmet
which shall contain a cheek pouch of sorts, to hold
a little writing-paper roll and a pencil, in the dry
air of the side of the helmet, at the left of my face.

It was the morning of April ninth when I went

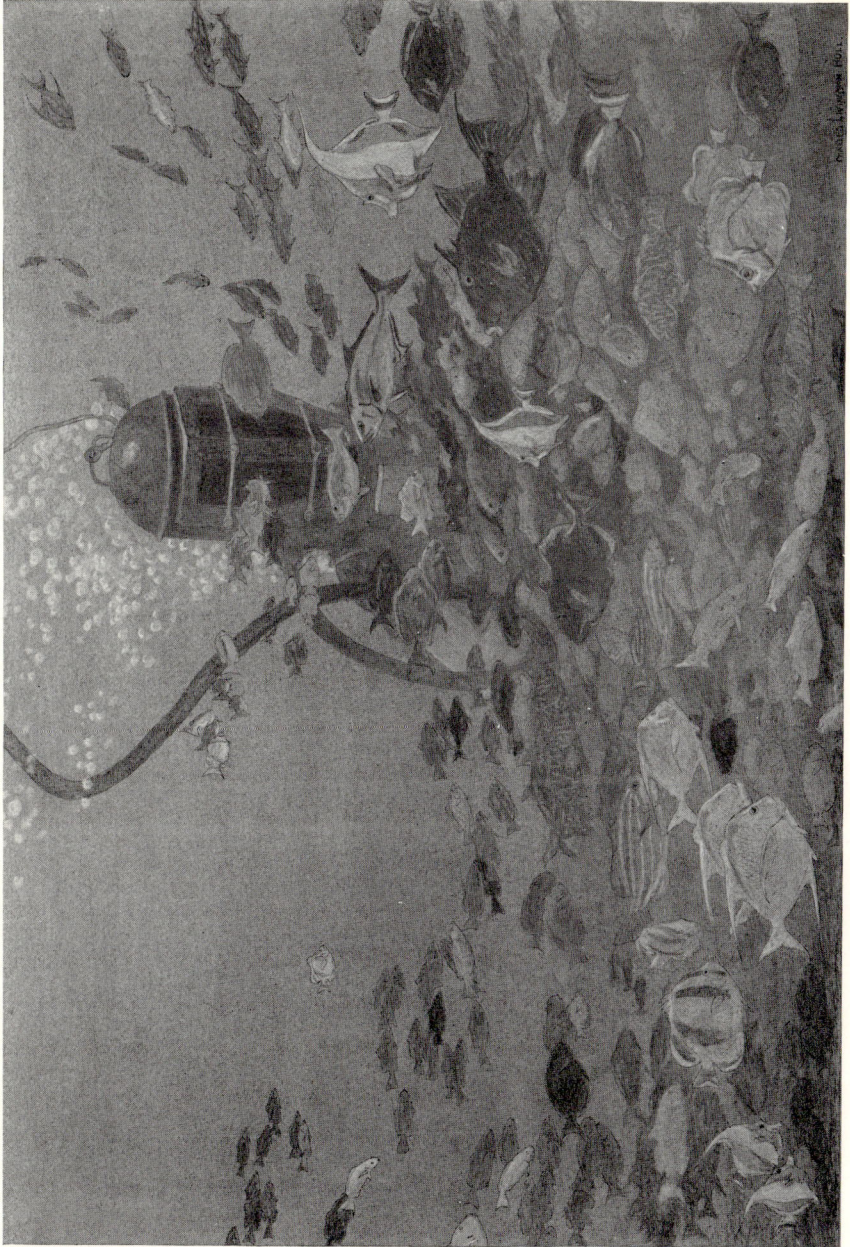

Fig. 18.—Diving in Twenty Feet of Water in Darwin Bay, Surrounded by Hundreds of Fish of Several Dozen Species.

FIG. 19.—MAKING READY TO DIVE.

Placing sixty pounds of copper helmet and lead weights over the head of the diver.

down for the first time, on a coral bank in Darwin Bay. I made five descents but recall very few details, because at the moment when I was ducking inside the helmet for the second time, I saw, a few yards away, one of the largest grey sharks I have ever seen, a giant of a generous eleven or twelve feet, cutting the water with his great dark fin. My companions did not fail to remind me of my notorious scorn of sharks, so with a rather sickly grin I went down. The dominant impression of this first experience was of the disconcertingly narrow field of vision—the oblique panes of glass in the helmet permitting only about sixty degrees. What I had seen at the surface kept my imagination busy with the keenest desire to see what was transpiring in the remaining three hundred degrees of my visual circle. I am certain that from above I must have looked like some strange sort of owl, whose head continually revolved first in one and then in the opposite direction.

It is idle to say that I, and I think all of us who went down, did not feel at first exceedingly nervous. It was disconcerting, as I have said, not to be able to see directly behind by a quick turn of the head, and until I became accustomed to the nibbling touch of some little fish who was investigating this strange creature so new to its world, I would often leap up in expectation of seeing some monster of the deep about to attack me. This stage passed and I soon felt perfectly at home. On the very few occasions when some creature seemed tempted to make a tentative hostile approach, it appeared to be the snaky hose extending to the surface and the constant stream of bubbles which deterred it.

In the afternoon of the same April day I submerged near the foot of the great cliffs, and, as I have described, disciplined myself into a greater realization of the wonder of it. I think my first surprise was of the constant movement of everything, not so much individually as of the whole in relation to the rocks and bottom. I knew of course that the boat was rising and falling with every surge, which heaved and settled in turn as each wave passed, to break against the cliffs. I found this same motion extended downward, with less and less force, until at thirty feet it all but died away. At present in about twenty feet of water I felt it strongly. I would be sitting quietly without the slightest tremor, when, gently and without shock, every fish in sight, every bit of weed or hydroid, the anchor rope, the shadow of the boat, the hose and myself swayed toward the land. One could resist it by clinging firmly to the rock, but the supreme joy, because of its impossibility in the air above, was to balance carefully and let oneself be wafted through space and deposited safely on the next rock. There followed a period of complete rest, and back again everything would come. It was so soothing, so rhythmical, that one yielded to it at times in a daze of sheer enjoyment. Where the water is not too deep and the bottom is sand or powdered shells, it is evident that the great surges are not a simple, compact movement, for here are made visible little, individual whirlwinds and casual, separate breezes which twist the shell-dust about or send up clouds of sand about my body.

In days to come I was to find the surge sometimes a very real danger, as when at Cocos I went down in a smashing thrashing sea and was scraped and torn back and forth across lacerating knife-points of coral and poisonous spines of urchins until flesh and blood could no longer stand it. Like getting one's sea-legs it soon became second nature to anticipate the swell, to lean against it, to shift the balance, so that everything moved except myself and the eternal rocks.

Now, day by day, occurred the accidents by which I learned how to do things, little by little relinquishing the ideas which, on dry land, had seemed feasible and important. For a day or two I could not understand why, during certain dives, the fish were so much tamer than at other times. The clue came to me when a rather heavy swell was running and I found that if I gave to the movement of the water, all the inhabitants, from gobies to groupers, from shrimps to sharks, accepted me as something new but harmless which the waves had washed in, but if I resisted the aquatic wind and maintained place and posture, I became an object of suspicion. This was the first of many radical differences which I was to find between the world of dry land and that of the under-water; on land, to move is to arouse fear among the wild creatures, here I did it by remaining still.

I walked or half-walked, half-floated, toward the cliffs. The rocks were almost bare in this bay, like those between tides, and the multitudes of lesser aquatic creatures were concealed beneath them. The water was quiet, and between surges was often perfectly clear, so that I could see plainly the cliffs rising high in air above that narrow straight line which marked the division between the two kingdoms. I went as far as my hose tether would permit and reached a boulder on which, the day before, at low tide, I had sat comfortably in the clear, cool air of the upper world.

Turning back, I saw that I had become a Pied Piper of sorts, leading a host of fish which followed in my train. The sun was out now in full strength and no fish, however strange and unknown to me, could hold my eyes from the marvel of distance. As I walked toward the cliffs I had also worked a little toward the east and the view I had, as I turned, was of another slope than that over which I had come. The bottom thus far was not wholly unlike the cliff above the water, but before me now the slope fell away in a manner which was beyond all experience—a breath-stopping fall, down which one could not topple headlong, but only roll and slide slowly, to be overcome, not by swift speed of descent or smashing blow, but by a far more terrible slow increase of pressure of the invisible medium, whose very surface film is death to us. To detect a faint, colorless shape now and then, through the azure curtain, and never to know whether it was rock or living creature—things such as this made every descent an ineradicable memory.

My range of vision was perhaps fifty feet in every direction, but for all I could tell it might have been fifty feet or fifty miles. The sun's rays filtered down as though through the most marvellous cathedral ever imagined—intangible, oblique rays which the eye could perceive but no lip describe. With distance, these became more and more luminous, more wondrously brilliant, until rocks died away in a veritable purple glory. No sunset, no mist on distant mountains that I have seen, could compare with this. One had to sit quietly and absorb these beauties before one could remember to be an ichthyologist.

As I was revelling in pure sensuous delight at this color of colors, a small object appeared in midwater close to my little glass window, and was instantly obscured by half a dozen little fish which darted about it, some actually flicking my helmet with their tails. Just as I saw that the suspended object was a baited hook, a baby scarlet snapper snatched at it, darted downward, and was at once drawn up into the boat. As I looked after it an idea came to me and I followed the snapper upward by way of the ladder. When the helmet was lifted off and I could speak, I expressed my wants, and descended again. Soon there fell slowly at my feet a small stone to which was tied a juicy and scarcely dead crab. I picked this up, waved it back and forth so as to scatter the impelling incense of its body and as if by magic, from behind me, from crevices upon which I was seated, seemingly materializing from the clear water, came fish and fish and fish. It is far from my intention to give a detailed list of all of these. The effect upon the reader in this connection, would be much the same as my own sensations at this time, if, by chance, my friend working the pump in the boat above had suddenly dropped off to sleep. Their names, numbers, colors and habits are all set down elsewhere in a more suitable place—*Zoologica*.

Even if I wished to speak of them in a homely way I could not, for most of them have had visited upon them the names only of the official, scientific census-taker, while the rest have no names at all. So Adam-like, I had to give them all temporary names, until I could identify them, or christen them with my own binomial terms. It was long before I could disentangle individual characteristics from the whirling mass (Fig. 18). The first four fishes rushed for the bait—

> "And yet another four;
>   And thick and fast they came at last,
>   And more, and more, and more—"

so that until I could shut my mind to the abstract marvel of it and my eyes to the kaleidoscopic, hypnotic effect, ichthyology gained little of specific factual contribution. I waved my magic crab, I may have murmured Plop! Glub! and Bloob! which is what the bubbles say when I first immerse—and the hosts came. Within three minutes from the time when the crab first fell into my hand, I had

five hundred fish swirling around my crab and
hand and head. Similes failed. I thought of the
hosts of yellow butterflies I have seen fluttering at
arm's length on Boom-boom Point; I thought of
the maze of wings of the pigeons of St. Mark's,
but no memory of the upper world was in place
here,—this was a wholly new thing.

Often there was a central nucleus a foot or
more in diameter, of solid fish, so that the bait and
my arm to the elbow were quite invisible. Twenty
or twenty-five species were represented, and, like
birds, they were graded with exquisite exactness
as to correlation of fear and size. The great majority were small, from two to four inches in length,
and these were wholly without fear, nibbling my
hand—passing between my fingers but always just
avoiding capture, no matter how quickly I shut my
fist. Six- and eight-inch fish also came near, but
were more ready to dart off at any sudden movement of mine. On the outskirts hung a fringe of
still larger fish, hungry, and rushing in now and
then for a snap at the delicious morsel which they
saw their lesser fellows enjoying, but always with
less abandon to the temptation of the moment. The
tameness of the little chaps, however, was so astounding, that the relatively greater wariness of
the larger fish scarcely deserved the name of suspicion, not to say fear. Another unexpected thing
was the rapidity with which these fish lost even this
slight suspicion and learned to connect my appearance with food. If I dived in the same spot several
times a day and several days in succession, fish

would approach in numbers and investigate my
hands and trident with much greater eagerness and,
I presume, with expectancy, than they ever dis-
played on the occasion of the first dive, before I
had repeatedly tempted them with freshly killed
crabs.   I could even recognize certain individuals,
characterized by some peculiarity of color or
form.

Before I go on to speak, even casually, of the
fish themselves, I must tell of my second discovery.
As with the crab baiting, and so much else in my
life, it was by sheer accident that I learned of the
possibility of spearing fish twenty to thirty feet under water.   The first few times I dived I carried a
powerful harpoon with a long metal handle, thinking I could lay it down and pick it up more readily
than if it had been buoyant. The big, green grouper
which I mentioned in my opening sentence was
bothering me, shoving his big jaws close to my arms
and legs, so I struck idly at him, missing of course,
and to my astonishment, he instantly attacked the
prongs of the trident.   Again I stabbed when he
was broadside on and struck him so hard that he
tore away with difficulty, whereupon he took
himself off, and sulked under a great mushroom
coral.

I remembered this incident and the following day
had a special grains made out of three large,
straightened fishhooks, fixed in the end of a yard-
long wooden handle.   This I took down with me
and waited until my regular crab bait came sailing down.   I caught the stone and wedged it in a

crevice of the rock, where the crab was only partly exposed. The fact of the invisibility of the food made little difference in the swiftness and the numbers of the arrivals. Their keen powers of scent drew them like filings to a magnet, and although only three or four fish could find room for a simultaneous nibble, yet scores waited behind, or pushed and wedged themselves in, reminding me of the buffet at a supper dance.

At last I decided to try my new weapon. On several former descents I had noticed a very common fish which was new to me, and now there were twenty or thirty in sight, nibbling at the crab, swimming in and out of crevices, and doing all the things which are imperative for small fish to do on occasions such as this. They were smug little fellows, high-backed like sunfish, brownish-black, with only two outstanding features,—delicately beautiful bright orange tips to the pectoral fins and a white base to the tail. Twice I leveled my trident and stabbed, and twice I missed. Then I found a new point of balance along the handle, struck again, and had a fish caught fast—my first *Pomacentrus leucorus* (Fig. 21).

And now my under-sea sprang a new surprise on me. Although I am a scientist and a hunting scientist, I hate to take life. Under the provocation of extreme danger to me or mine, I have always valued human life at less than nothing, but shooting down a savage as he is rushing you is one thing and deliberately spearing a fish which you have been watching and which swims about close to

your face and hands in perfect fearlessness is quite
another. However, one can be tender-hearted without being sentimental and if I need the facts for
science, to complete the life-history of a whole
species, I will shoot a dove on her eggs without
compunction. I sympathize, on the other hand,
with the Hindoo fishermen of the Laccadives who
are not allowed by their faith to take life, and
hence, when they have drawn their nets, they rush
ashore and lay the still living fish gently upon
leaves and moss. Later they return, and finding,
to their surprise, a lot of fish which are quite dead,
it is permitted that they gather them up to sell or
to eat.

So it was not with the unmixed feelings of a tri-
umphant Neptune or a successful ichthyologist
that I clambered up the ladder, and when near the
surface held out my trident with the impaled fish.
My pleasure in the feat was heightened when I
finally ascended and found my fish swimming unconcernedly about in the well of the boat. As a
matter of fact, a much greater percentage of my
speared individuals recovered and survived, living
and feeding contentedly for weeks in our aquaria,
than of those we caught on hook and line. Almost
invariably the tip of the grains would penetrate
only the mass of back muscles, leaving quite untouched the head and the vital organs of the body.

I experimented with all sorts of methods, such
as putting a bit of crab on the trident itself. This
was a complete failure, for the fish would crowd
around it head-on, and with all my efforts I never

FIG. 20.—WILLIAM BEEBE IN THE DIVING HELMET.

It merely rests on the shoulders, kept there by its own weight.

Fig. 21.—The Author in Twenty Feet of Water, on Large Coral Boulders, about to Harpoon a Rare Fish.

Fig. 22.—Moorish Idol, a Brilliant Shore Fish of the Galápagos.

succeeded in even touching a fish when in this position. It can very naturally shoot forward and backward with infinitely greater speed and facility, than
move sideways against such a heavy medium. So
my efforts were always directed at fish broadside
on. This method of attack was so new to their experience, that even when just missed, they darted
aside only far enough to escape the thrust, then
returned at once and examined the trident with
deep interest. Sometimes I would scrape off a few
scales and then these most astounding creatures
would rush back in great excitement, and snap up,
one by one, each floating scale, "getting a bit of
their own back," as it were.

The smaller fish were as easy to reach with the
prongs as if they were blackberries fastened to a
stem, but they were so small and agile that they
slipped between and around the barbs. The easiest
of all to secure were the medium-sized herbivorous
fish such as the yellow-tailed surgeons and the
gorgeously colored angelfish. These came inspired
only by curiosity and drifted about me aimlessly or
nibbled at the rock by my elbow. The sign of
Cancer meant nothing to them, and their efficient
poisonous spines or defence of whatever kind
wrought a self-confidence which carried them
through life calmly and without fear. I had merely
to wait until they approached and turned their
broad profiles when a quick flick of the wrist meant
their transference to life in one of our aquariums—
where they continued to live placidly and undisturbed by any change which fate had brought to

them. The number of the surgeons which I took
was limited only by my desire for specimens or the
capacity of our aquariums, for my capture of one
conveyed no alarm or sense of insecurity, and when
I again climbed down the ladder the chances were
that I would find the remainder of the school in the
same spot, undisturbed.

The best sport was to be had with the brilliantly
colored wrasse. They were among the most active
and swift, slender and supple as eels, with an abundance of fins for doing everything that perfect control demands. Two species in particular were
always about, although never more than a half
dozen were in sight at once. Nature must have
relegated the coloring of some of these fish
to an amateur assistant, for it was crude, blatant
and, judged by human ideas of ornamentation, in
execrably bad taste. Yet as I saw it—a living
organism—winding in and out of dark crevices, or
twisting almost on its back to get a nibble at crab
meat, it seemed rather an exquisite mass of palette
splashes. The head was scarlet, the body, fins and
tail mostly bright grass green. The head was outlined in dark blue, and from the lips, which were
solidly of the same color, five blue lines streamed
backward, flowing in irregular bands through the
eye and across the cheeks, saturating the pectoral
fins. The whole green body was thickly banded
with irregular vertical lines of an unnamable dull
maroon—like thick heavy streaks of some awful
rain or acid stains. The tail had a stiff, unnatural
pattern, like a great scarlet H drawn crudely over

the green. I was happy when at last I outwitted a six-inch green wrasse, and put him aboard, where he lived for two months, allowing us to paint and study him at our leisure.

The other wrasse was simpler, but even more striking in pattern and coloration, and to the last defied my every effort. Twice I struck and marked them, and day after day the same individuals would come about as bold as ever, flaunting their scars and wounds in my face. One of these had two jagged holes well into his side, yet they apparently gave him no concern, nor interfered at all with his speed and control, and he easily avoided every attack which I launched. These fish were about five inches in length, bright tyrian purple over all, with a broad vertical band of sulphur yellow extending down from the neck around the body and including the pectoral fins. While I was exerting every muscle to get him, I called him many names in the quiet of my helmet, but these are neither here nor there. No written description fits him, and until I return and with greater skill succeed in overcoming his cleverness, he can be called only the Yellow-banded Purple Wrasse.

Armed with my crab meat I often sat and watched with my face close to the center of interest. The mass of fish was composed of a bewildering array of forms, yet both here and in the other islands where I dived, there were several dominant species. On almost my first dive I welcomed with a shout—a shout which echoed only within my cubic foot of air—my old friend of two years ago, the blue-lined golden snapper, *Evoplites viridis,* which I then pictured in color.[1]   These beautiful fish were abundant, and although many quickly gathered when crab meat was provided, yet as a rule, they were solitary, swimming about singly close to the bottom.   One day in one spot we caught thirty-eight with hook and line as fast as we could pull them in, but none of these lived, while one which I harpooned thrived for many weeks.   Most were six- to eight-inch fish, but occasionally I caught fleeting glimpses, in deep water ways, of giants nearly three feet long.   They were voracious and when they dashed in for a snap at the crab, they often seized the entire joint of a leg which they swallowed whole.

The little round, brownish-black *Pomacentrus* fish of two species were the most abundant of the four-inchers, and, as I shall relate more particularly in another chapter, were the most absolute home bodies, each living in his particular crack or crevice, from which he frequently rushed out and attacked ferociously any fish which approached too near, regardless of its size.

Another field of work of tremendous interest was suggested when I turned over the first stone and saw the mass of life covering the underside and filling the crevices.   I arranged to have a pail lowered on a rope, and squatting low on the floor of the bay I filled the pail and gave the signal to draw it up.   Five pailfuls provided a tub of rocks. This was left standing in the sun for a day and at

[1] Galápagos: World's End, Plate V.

the end of that time there had crept out an amazing array of interesting beings,—beautiful sea-worms, starfishes, squillas, hermit crabs, and shrimps of every hue, a number of strange larval fish and an adult formed, wonderfully patterned, quite fearless moray eel exactly one and one half inches in length. This tapped a fertile and untouched field, providing organisms which cannot be dredged because of their shelter under and within coral and stones, and not to be gathered by wading along shore at low tide, since twenty feet of water lay above them.

The obliquity of the two windows in the helmet made it necessary to look out of either one or the other exclusively, when engaged in observation or work which required accurate correlation of eye and hand. Seldom have I seen a funnier sight than the earnest efforts of any of our party before they learned of this optical effect. Through the water-glass a pale figure would be seen crouched on the bottom, industriously picking up stones and carefully dropping them about two feet from the bucket. After much hard labor, the helmeted creature would raise the empty bucket and gaze at it in puzzled astonishment. In imagination we could see the large question mark poised in mid-water over his head. Another labor-saving individual decided to pick the specimens themselves off the rocks, and long streamers of algae and clumps of hydroids were gathered and carefully placed in the bucket, only to float instantly out and up to us, while he was looking for other equally buoyant specimens. Don Quixote's horse was nothing compared to the worker's ultimate idea of the capacity of that pail (Fig. 20).

From first to last I could never guess, from examining the bottom through a water-glass, what a submersion would yield, or even look like, except in the most general, superficial way. It was like judging a shore line from a ship with all the indentations flattened, all the coves and little bays concealed in the optical straightening, and the wicked, crashing breakers smoothed from behind into harmless appearing ripples. In many lights, the bottom, even only twenty feet down, appears merely undulating or paved with huge stones.

One of the last dives I made in Darwin Bay showed such an aspect from above. I went down rather deeply, but very slowly, for I always came under the spell of the ever wonderful blueness of distance. It seemed impossible, even after all the times I had studied it, that invisibility or opacity of whatever distance could result from such a luminous medium. When at last I rested on the bottom I watched three white-striped angelfish chasing one another in sheer play. They drew my attention upward to where they were breaking the surface film, not far from the boat whose keel was bobbing absurdly up and down. The angelfish then curved downward, the long filaments streaming from the fins above and below, and giving the appearance of even greater speed. They rose and fell, circled about, turned on their backs and fell into nose dives as easily as I sat still. Finally, the emotion over, whatever it was, they all came to rest still high up in mid-water. It occurred to me that in comparison, our own world is practically one of two planes, while this is really the one of three. It is fair to compare fish only with birds, and even birds need two perching props, and do not dare to develop wings or feather fins beneath the body, for, sooner or later, they must alight, while a fish can live, eat and sleep poised in mid-water.

I turned my attention from the fish to the scene behind me and the absurdity of my appraisement from the water-glass became apparent. I was standing a few yards away from a boulder as big as a cottage, and my heart gave a leap as I saw a curved flight of steps—giant steps like those up which I had once climbed Cheops. They began on my side at the doorless entrance of the sinister cottage, slowly encircled it and vanished behind it in a soul-stirring abyss of blueness, which, from a delicate shade near at hand, blued more and more clearly into infinite depth and space. I believe that Sime would have loved this scene, and Dunsany would have deemed it not unfitting for the habitation of Gnoles. Töten Insel treasured no more mystery in its perspective than did this. As I watched, a bit of greenish black coral which projected eave-like, began to move and crawl slowly downward, and with it went dangling things which I had taken for strands of dead seaweed, but which on this edifice might well have been awful stalactites or icicles of sorts. The octopus climbed down, hesitated, felt about in different directions, and then descended the steps, flowing along the angles like some horrid viscid fluid in animal form. The most active imagination could not have set the scene better, or found a more appropriate actor.

But like the double miracle of the stars falling into the volcano the end was not yet. A mist of yellow-tailed surgeons drifted across the stairs and the dread boulder, and for a moment their calm matter-of-factness lessened the sinister feeling of the whole thing. A strong desire arose to look around the corner of the stair for myself. I was submerged so deeply that as I stood, I could barely reach the lowest rung of the ladder, indeed I was occasionally lifted a few inches from the ground as the boat rose to a greater swell. But I knew the hose was new and stout and even if I began to fall with that terrible slowness, as seemed easily possible to my imagination, I could surely climb back up my own string. One finger relaxed and I was about to take the chance when a mote, very faint and pale, stirred the blueness as if some wondrous tapestry curtain were troubled by a breath of air.

The thing grew denser, took form and became concrete, and a flat, round-fronted head, lazily undulating, wound through the water over the steps, a nine-foot shark weaving along where I would have been a minute later. My common-sense theory of the harmlessness of these beings still held good; in the last few days dozens of them had approached within a few yards of me, but the eerie character of this place had penetrated even my prison of copper and glass, and when I realized where my precious ladder would drift to when I relinquished my hold, looked down at my unprotected limbs and realized that I had not even a trident with me, I decided to go through life with the mystery of the stairway unsolved. The great, grey being, wafting along its hundreds of pounds of body by slow, gentle undulations, kept on and on until again hidden by the blue light. When I ascended to a world of greater reality, I took with me the memory of the beings to which legend and fact have brought the greatest notoriety of anything in the sea, and the setting in which I found them will never pass from mind—the Edge of the Edge of the World.

CHAPTER IV

ALBATROSSES

THE first time I ever saw an albatross was at dawn far out in the Indian Ocean. It was that hour at sea when perspective does not exist, and, like the houses of a tropical coastal city, everything appears flat and on one plane. I was observing a small flock of petrels from the rail of my vessel when a lighter colored bird appeared above them, apparently of the same size. As I watched, it grew larger and larger, until, to my amazement, it joined the petrels, and in the same instant they were dwarfed to insect size while this white bird assumed relatively gigantic proportions, and I knew that I was seeing the effortless flight of an albatross.

For years thereafter my eyes were always on the lookout for these birds. In southern seas and in the north Pacific one may hope to find them, but not on our own boreal Atlantic. A great many years ago, however, long before man began to have sufficient perspective of his ancestry to worry about it, albatrosses were calmly winging their way over our northern seas, and we find their fossil bones both in England and America. A vast amount has been written about their flight but today we watch them with quite as much ignorance of how they contrive it as when the first mariner saw and marvelled. So close to the water they skim, so automatically they rise and fall, outlining the unpredictable movements of waves, that they seem to possess all the secrets of white shadows. When we watch closely and less emotionally we begin to see the part which wind plays in the support of this relatively heavy mass of flesh and feathers, throughout the tens of thousands of its miles of progress. The albatross is never so supreme and relaxed and effortless as when it is coasting up-wind, but a breeze on the quarter is less sustaining, and when flying with the wind, frequent circles and intersecting spirals are necessary to attain and sustain sufficient impetus and altitude. This is the fame of the bird, and throughout history and literature almost every mention of it has been synonymous with supremacy in flight.

Once seen and recognized, an albatross can never again be mistaken for any other bird; its great size, the unusual length and ribbon-like narrowness of its wings, the large, yellow, hooked beak —all these mark it even at a distance. The ease and lack of effort of its flight are deceiving, and only when it circles and encircles a fast-moving steamer do we realize the terrific speed of which it is capable.

Albatrosses are usually classified as a family in the order of birds known as Procellariiformes, or oceanic swimming birds with the nostrils arranged in two long tubes lying along the beak. Their nearest relatives are the hosts of little black and white petrels or Mother Cary's chickens which abound on every ocean and are familiar in storm and calm. In fact it would not be far from the truth were we to call petrels dwarf albatrosses, or the latter giant petrels. Diversity in size is probably as great in this group of birds as in any corresponding assemblage of animals on the earth. Within sight of one another I have collected an albatross and a petrel, the former weighing one hundred and fifty times as much as the latter, while the albatross had a spread of wing seven times as great as that of its tiny relative. There has been much written of truth and of exaggeration in regard to the wing spread of albatrosses. I am inclined to agree with the words of Dr. Lucas, who writes of the wandering albatross "it is also the largest species, having a stretch of wings of about twelve feet—an assigned dimension of seventeen and a half feet being either a great exaggeration or highly exceptional." In the Eocene, however, there lived an albatross-like bird, which, judged by the size of its fossil bones, must have had a spread of wing of at least twenty-two feet.

In birds so evidently related as petrels and albatrosses but differing so greatly in actual size we have most interesting evidence of possibilities of flight character. It would seem impossible for any small bird to soar for any length of time or to go for any distance without actually flapping. I can recall no bird of small size which has this ability, while such past-masters of non-flapping flight as

vultures, pelicans, screamers and albatrosses are
all large and heavy of body. I have made over
three hundred flights in airplanes myself, in peace
and war, close to the ground and once up to an
altitude of twenty-two thousand feet, yet the way
of an eagle in the air is still, to me, inimitable, and
always will be unless we can duplicate its great air
chambers, the lightness and strength of its hollow
bones, and the friction-evading plumage.

The part which albatrosses have played in relation to man is interesting. First, admiration for
its flight by early mariners, and a sense of compan-
ionship and camaraderie in its society in the deso-
lateness of mid-ocean.

> "And a good south wind sprung up behind;
> The Albatross did follow,
> And every day, for food or play,
> Came to the mariners' hollo!"

This feeling, in the course of years, very naturally developed into an affection, and this, vitalized
by the superstitious sub-stratum of the seaman's
mind, increased to a resentment of any attempted
injury.

> "God save thee, ancient Mariner,
> From the fiends that plague thee thus!—
> Why look'st thou so?" "With my crossbow
> I shot the Albatross."
>
> And I had done a hellish thing,
> And it would work 'em woe:

For all averred I had killed the bird
That made the breeze to blow.
"Ah wretch!" said they, "the bird to slay,
That made the breeze to blow!"

The extreme of ridiculous theory is to be found
in a very old book by Wiquefort, who says "thefe
birds are often feen fleeping in the air, entirely re-
mote from land, with their head under one wing,
and the other employed in beating the air!"

There came a day when the homes of these birds
were discovered, usually a tiny coral focus of the
scattered individuals which roam so far and wide
over the oceans. One island became known to some
Japanese who had neither pity nor superstitions,
and before President Roosevelt could enforce his
sanctuary legislation they had starved or carved
alive nearly a million albatrosses for their wing
feathers which were sold to milliners as eagle
plumes. Then sentiment and kindness again became dominant—the feather markets in our cities
were closed and wardens appointed on the tiny
islets, and if the desire which museums have for
endless series of skins can be controlled, it may be
that for many years these magnificent birds will
continue to share this good earth with us.

There is an authentic record of an invaluable,
although it must be admitted involuntary, benefit
rendered to man by an albatross. Some years ago
there fell exhausted and dying from starvation
upon the beach at Freemantle, West Australia, a
great albatross. When found, it had a tin plate

fastened around its neck on which was scratched the news of the wrecking of the French ship *Tamaris* three weeks before, and the survival of thirteen of her crew on Crozet Island. During this period the albatross had flown over four thousand miles of ocean, too terrified by its burden to stop to feed. It was a remarkable incident, quite reversing the experience of the ancient mariner;

"Instead of the cross, the Albatross
About my neck was hung."

Intellectually, man's relation with albatrosses has been less spectacular but of equal interest. Linnæus, one hundred and sixty-eight years ago, first played taxonomic Adam to the albatross, calling it *Diomedea exulans*. Its godfather was probably therefore the famous hero of the siege of Troy, but Grecian etymology provides a much more poetic and appropriate derivation, and it is pleasant to think of the albatross, whether winging over foam crests or at home on its little isle as being ever *Dio-medea* or God-counseled. In its specific appelation Linnæus was also happy for to the ordinary observer, the wandering albatross is truly *exulans*—homeless, banished apparently from all connection with solid land.

It must be confessed, however, that Linnæus made a *faux pas* when he was led to associate in the genus of the great-winged albatross, the little fin-winged penguin—birds as unlike in habits as they are in physical makeup—suffering comparison only in their astonishing divorce from land,

and their extreme adaptations for continued existence in the air and the water respectively. One
can readily tell that Linnæus was a closet, or at
least a terrestrial, rather than a sea-going scientist,
for his contribution to the habits of the wandering
albatross are, *"æthera altissime scandens; victitans
e Triglis volitantibus a Coryphæna exagitatis."*
But this bird neither attains great altitudes in the
air, nor does it, to my knowledge, capture living
flyingfish.

In more recent years, as additional specimens
have been secured, more and more species have
been differentiated, until the family Diomedeidæ
now contains several genera, about a dozen well-
marked species and perhaps twice as many more or
less clearly defined subspecies. These latter dis-
tinctions must always be checked up with the fact
that there are several distinct changes of plumage
from nestling to adult, while, like most birds of
large size, albatrosses develop rather slowly, and
in addition to this there appears to be an unusual
amount of variation in birds of the same colony and
age.

But I shall get away from the spirit of this
volume if I do not return to the living birds themselves. Fortunate is that bird or animal on the
earth today which has found an isolated niche for
itself, where it may claim comparative sanctuary.
And this does not necessarily mean isolation from
a geographical point of view. It may be a gastronomic one, such as the scavenger vultures have
achieved, or the tough leaf diet of the hoatzin,

feeding on substances which are disdained by their
fellows. Or it may be an isolation from fear of
death by daylight, such as is engendered in bats and
goatsuckers; or from actual low development of
mentality as in the sloth; or an optical sanctuary
such as an insect which in color, form and movement strives ever to be thought a leaf. But no
more dramatic isolation exists than that of the albatross, which, although furnished with legs and
toes, yet for most of its days spurns all solid earth
and lives its life between sky and sea.

When I first saw albatrosses at their breeding
ground I experienced a slight feeling of embarrass-
ment, as if I were peeking through the blinds, or
looking behind the scenes. I feel much the same
when, in the rotagravure section of the Sunday
paper, I see a photograph of some famous prima
donna making an apple pie in her kitchenette.
The voice of a *chanteuse* and the flight of an albatross are among the more wonderful things in the
world, so much so that at first we hesitate even
to think of the authors in relation to the trivial
things of life. Whatever may be the case with the
home life of a great singer, that of these famous
birds shows the inevitable law of compensation. I
have already courted displeasure in revealing an
unromantic side of the Sargasso Sea, so I might
as well continue and describe the gait of the Galápagos albatross. Its progress on land makes that
of Charlie Chaplin appear grace itself, but for
sheer amazing interest, the courtship and dances
of these birds vie with its flight.

It was late in April when I completed my vain search for the Humboldt Current south of the Galápagos. After making certain that there were no signs of it within a radius of at least one hundred miles south of the Archipelago, I decided to steam north, and sighting Hood Island, anchored in a beautiful bay on the northwest side. As the *Arcturus* slowly felt her way toward the shore a flock of large black birds, swimming in single file, appeared off the port side and puzzled me until my glasses showed them to be albatrosses. Next to the active volcano on Albemarle this was the most exciting thing that Galápagos had offered on this trip. A few days after we had begun work at Hood my scouts reported that they had located the rookery several miles to the eastward. So on the evening of the twenty-fifth of April I made my preparations to visit them next day. That evening will not be forgotten by any of us for it was then that all the giant flyingfish in the world came alongside.

As usual, after dark, I lowered a cluster of electric lights near the water, and several of us took our places on the steps of the gangway. Almost at once large flyingfish began coming, and we caught several hundred. Looking down on them through the water their bodies appeared a beautiful pale green, and the wings bright pink, but in reality they were steel blue and wine color. About one in every fifty had the wings densely covered with round black spots. Never have I seen such uninterrupted terror or constant fear. During all the early part of the evening we could see nothing in pursuit, no hint of enemies, yet the flyingfish dashed frantically about within the blaze of light, quite heedless of where they were headed. We had to be on our guard, for they would strike with terrific force against our bodies and the side of the *Arcturus*. Often such a blow against the ship would knock them senseless. Several rose and passed between the fourth and fifth steps of the gangway, so high were they able to rise from the water. Now and then there materialized a real reason for their terror, as a sea-lion shot into view, seized a fish and vanished. Far down we could see the forms of sharks—ghostly pale in the dim light —but none came close, except once, when I saw a shark rise and engulf one of the disabled flyingfish. These flopped by twos and threes into the small boats and in the morning we picked up over a hundred. It was an astonishing and a memorable sight, few of the fish being under a foot in length, and in such endless numbers.

Early next morning we started in the *Pawnee* launch eastward along the coast between Hood and Gardner Islands. We passed close to wonderful sea caves and could hear the swells booming into spray far within. At the entrance blue-footed boobies and pelicans perched motionless like guardian gargoyles and every available niche and ledge was filled with nesting noddy terns.

Our course took us head-on into the great rollers and as I lay flat on the forward deck, the walls of water ahead looked like solid jade,—the early morning light shining through the thin emerald tops of the swells and tingeing our boat with the same color. Dolphins now and then rushed at us and leaped so high into the air that I could see the rising sun beneath their curving bodies, drenching me with spray as they fell back.

Four miles beyond we drew close inshore, but had to pick our landing place with much care, and even in the partly sheltered beach which we found, the cameras and guns barely escaped a soaking, as we staggered ashore up to our arms in water. A walk of half a mile inland took us around an impossible bit of shore, through a growth so dense that we could see only a few feet in any direction. Everywhere we passed the dry beds of drainage streams, through which, after cloud-bursts, torrents must make their way to a lake which at present was a half-dried marsh. This was filled with mosquitoes, apus and branchipus fairy shrimps. Much of this end of the island was almost level, with occasional tall pinnacles or spires of broken lava which rose high and slenderly above the scraggly trees and shrubs. The going was easy as there was only a small amount of thorny mes-quite and cactus. Black finches and mockingbirds chirped and flew close to us, and rarely a great bee zoomed past.

Suddenly in front of me a white head and neck shot up from behind a bush, and a few steps brought me to my first nesting albatross. After this we found them scattered about, sometimes singly or again in pairs, with or without an **egg**.

Fig. 23.—Hood Island Rookery of Boobies, Gulls and Albatrosses.

FIG. 24.—GALÁPAGOS ALBATROSS ON ITS EGG.

FIG. 25.—THESE ALBATROSSES HAVE NO FEAR OF MAN.

Twice I came across an unattended egg lying absolutely in the open on the level, red lava soil, without
the slightest hint of nest. Not far away were six
pairs of birds sitting close together.

Coming over a low rise of ground we suddenly
saw the shore close at hand, and a most wonderful
panorama to east and west. Two headlands curved
around before us to the right, while straight ahead
a third ended in a high arched natural bridge of
lava. Everywhere, from our feet to the tip of
the headlands, were nesting birds—thousands of
pairs of olive-footed white boobies, with small
colonies of frigatebirds here and there, and occasionally a pair of blue-foots. Galápagos gulls were
sitting on their eggs beneath the arch of the bridge,
and shearwaters swooped in and out of the foam.
The boobies were well along in their nesting season,
for the ground was covered with half-grown,
snowy-white nestlings, which unceasingly snapped
and squawked at us (Fig. 23).

Back of the headlands and all along the shore,
somewhat removed from the main mass of nesting
birds, were the scattered albatrosses, probably a
thousand all told, two or three pairs close together,
or a single bird quite isolated. Some were casually
resting, and these rose to their feet at my approach
and waddled slowly off. But most had already
chosen their nesting site and refused to leave either
the bare eggless space upon which they squatted,
or the great oval shell which they kept so close
beneath them. The difference between the albatrosses and the other breeding birds, in respect to

my presence, was very striking. The former watched my approach gravely and without fuss or sound chose their course of action. If on an egg they permitted no familiarity, but snapped with their powerful hooked beaks, and vigorously resented any advance. With a stick I gently pushed one of the great birds back until the egg was uncovered, then took it up, examined it, and replaced it, when the parent, with no show of resentment or worry, shifted slowly forward, opened wide her breast feathers and gently sank close down upon it again (Fig. 25).

I am describing this rookery of albatrosses calmly, as if it was to me merely an extension of the myriads of nests of the other seabirds. But in reality it was one of the great experiences of my life, set apart from the rest of the rookery as Buckingham Palace is from the houses of Grosvenor Square. Here at last was the bit of dry land where these splendid creatures of the air deigned to alight and to carry on the affairs of everyday life.

I saw one coming in from the open sea, steadily as a triplane, without quiver or shift or balance of wing. When over the level ground the wings were tipped backward—the under surface presented as a brake, the legs lowered, the head held up, and with all its might the albatross bore back and began paddling furiously with its great webbed feet, seeking foothold as it taxied over the rough ground. Slower and slower became its speed, and finally the wings half reefed and gave

up their power. But the feeling of land was too
unaccustomed a thing—the bird sagged sidewise,
tipped over a pebble, half fell across one of its
fellows, and turned over, rolling undignifiedly
several times before it quite stopped. Then it
rose unsteadily, gathered itself together and looked
around, clattering its beak and shaking its head,
doubtless, saying to itself, that the land was not
what it used to be.

I watched this bird and followed it for a considerable distance inland, but at its very first step I
realized anew how far specialization for the air
had gone. Flat feet, fallen arches, rheumatic joints,
crippled limbs—all were suggested in its painful,
appallingly awkward gait. At each step the entire
body turned with the leg, and the whole head and
neck swung around and down on the opposite side
to aid in balance and in supreme endeavor for each
succeeding step. I have never seen a more un-
gainly, effortful mode of progression, and when
thrown on the motion picture screen it arouses as
much amusement in an audience as the peripatetic progress of Charlie Chaplin. Some day an
epic will be written on the law of compensation,
the most dramatic thing in nature—the peacock
with its aristocratic, incomparable display of exquisite colors, and its Billingsgate squawk of a
voice; the nightingale, embodiment of glorious
soul-stirring song, with feathers of dullest russet
and grey. And here were albatrosses, master
flyers, tottering miserably along as if each step
brought acute agony.

I walked slowly after my particular bird, sitting down to rest when it sat down, and trying to keep from laughing aloud at its frantic efforts to surmount the least inequality in the ground. A hundred yards were traversed before it came in sight of another bird which seemed to be its mate, resting upon an eggless bit of volcanic gravel. Before the birds met, mine sank exhaustedly upon the ground, ostrich-fashion, and settled down for a rest. I squatted about ten feet away and realized for the first time the real beauty of these birds. The great hooked beak is golden yellow and the head and neck purest white; the entire body is freckled with wavy grey, and the mighty wings are dark brown. Over the eye the feathers beetle outward in a most curious vizor-like fashion, and the large, dark eyes give the bird a gentle, kindly expression. Measurements taken later on showed that these albatrosses averaged over three feet in total length, and eight to nine feet in extent of wing, while the weight was about ten pounds.

As my particular albatross seemed settled for a while, I called on another bird about twenty feet away and gently pushed her off her egg. This I took for science' sake, salving my conscience with the certain knowledge that it would soon be replaced with another. It was a beautiful thing, as indeed all eggs are, very broadly oval with blunt, rounded ends, and measuring about three by four and a half inches. It was white with a dense cap of deep reddish brown at the larger end. This color died out in a sparse speckling along the sides,

leaving two-thirds of the shell immaculate. I later
found that the contents made a delicious omelet
after being carefully extracted through the tiniest
of holes. I clearly remember the thrill when I blew
my first egg, that of an English sparrow thieved
from a mass of hay behind the attic window of my
home, and now as I held this great shell in my hand
I thought of all the eggs I had seen between,—
those of pheasants in Tibet, junglefowl in Java,
hoatzins in Guiana and hawks on the summit of
Cheops, and I was grateful that the first thrill had
in no wise lessened (Fig. 24).

But my bird showed signs of continuing its
promenade, so I hastily returned. A half hour
later as I was covering my last bit of paper with
frantically scrawled notes, it occurred to me that
there are three phases in the life of an albatross,
each of which arouses in us a widely different, but
profound emotion; first, admiration for its superb
powers of flight, second, amusement at its ridiculous gait, and third, sheer amazement at the elaborate detail, studied sequence and formality of its
courtship and play.

I had read accounts of this at other rookeries of
albatrosses, but no description prepares one for
the actual performance. My bird walked up to its
mate which, in its turn, rose and faced the new
arrival. They stood with their breasts about a
foot apart. My albatross suddenly shot its head
and neck straight up, the bill pointing skyward,
uttering at the same time a deep, grunting moan.
Its partner followed suit, then, alternately, each

bird bowed deeply and quickly three times. Without an instant's delay they next crossed bills and with quick, vibrating movements of the head, they fenced—there is absolutely no other word for it —with closed mandibles. Without warning my bird ceased and again shot his head high up into the air. Its mate instantly turned her head and neck far sideways and held them motionless and concealed from my point of view, close to the left wing and side. Then another double bow and a second bout. Next, both birds rested, looking quietly around as though nothing unusual were in progress, when the mate gave the stretching cue in her turn, and there followed a long bout of the fencing, this time my bird with widely opened mandibles, the other's beak even entering its mouth once or twice. For five minutes this performance kept up, when a third bird approached, bowed and engaged my albatross. This was only half-hearted however, and the third individual soon waddled painfully away, and the first two resumed the astonishing ritual.

I walked over to the third bird and bowed deeply and to my delight it bowed in return. Seeing no rapier bill, however, it solemnly walked away, until I again faced it and bowed when it returned my salutation twice and took a step toward me. That, alas, was as near as we could come to an engagement, but I shall never forget my amenities with this feathered D'Artagnan of Hood Island.

This intricate performance has been described

before, and probably all albatrosses go through
something of the kind. It is *au fond* unquestionably a courtship, but I really think that it provides
some sort of pleasure other than this alone, for
I have seen it indulged in by a pair of birds which
already possessed an egg, and again I saw one individual go through part of it with at least three
other albatrosses in turn.

The sequence is not always the same, but the
upward stretch always begins it, and all the phases
are enacted by each bird in turn. The grunts or
groans or rasping notes are sometimes frequent or
the whole thing may take place in silence. There
is no emotional climax. It begins and ends in the
calmness which the gentle eyes of these birds and
their philosophical treatment of an intruder such
as myself indicate as a deep-seated character.
Fortunately it was a very easy matter to obtain a
perfect series of motion pictures of the fencing,
and thus to preserve what mere words so completely fail to delineate.

I found that the sole food of these albatrosses
consisted of rather small squids, and this seems to
be the case with all other species, although like
gulls, they have learned that galley scraps cast
overboard are delectable dainties, and will follow
vessels for many hundreds of miles on the lookout
for such manna. As a quite radical change from
squid diet there was once found in the stomach of
a wandering albatross an undigested Roman
Catholic tract with a portrait of Cardinal Vaughan.

When we finally left the rookery we walked up

to two isolated, lonely birds, gathered them under our arms and took them back with us. At first, on the *Arcturus*, they suffered severely from sea-sickness, but when they recovered they ate freely and lived for months in the Zoological Park in New York, admired by a host of people whose only acquaintance with an albatross heretofore had been through the woes of the Ancient Mariner.

CHAPTER V

IF there had never been but one opal, one peacock, one sunset and one butterfly what glory of
history and legend would accrue to each. Men
would have sworn great oaths of promise upon them
and made them into sagas, they would be the ultra
similes, a religion might have been founded upon
one. But opals are worn for unlovely reasons upon
unlovely hands, the man-given name is often
deemed more important than the god-given idea of
a butterfly, and a sunset, if not less than an inter-
ruption of dinner, is slighted because of the certainty of another on the following evening.

When a very wonderful thing comes into our
lives for the first and perhaps last time, we betray
our very birthright if we do not meet it with all
the feeling and emotion and intellectual appreciation which is our human prerogative.

The *Arcturus* was anchored close under the steep
cliffs of Darwin Bay—much too close for the Captain's ease of mind. In late afternoon I leaned
on the rail and watched the gigantic blocks of
basalt catch and reflect the salmon and coppery

bronze of the sunset behind me. Now and then it seemed almost as if they answered with a faint tinge of their own. I had spent much of the day in a diving helmet, clambering far down about the lower reaches of these same cliffs,—in a world where the word dry is without meaning and where the shadows of sharks instead of frigatebirds follow as they pass over my head. Now, for a while, the birds, even the garrulous gulls, were silent, and the quiet of my ears lent more power to my eyes which scanned the aged cliffs. So silent, so dead, so hard, so immutable were they, that no continent seemed more permanent than this tiny islet in mid-sea. Surely, when the world first cleared its face, these eternal iron cinders were here—cold, motionless, black as night.

There was still a tinge of rich color in the west when I went below, to eat and work, and talk of things so unimportant to stars. It was long past midnight when the booming voice of the Second Mate broke into my dreamless sleep, and brought me on the instant to my feet, clear-thinking and listening,—the heritage of a myriad nights in jungles and deserts. He beckoned me to the bridge and pointed toward the entrance of the bay. Something was there which should not have been. If at this hour, on the equator, a sunset afterglow still lingered, then indeed had the stars turned backward in their courses. A sudden idea came to me: "A ship on fire?" The mate shook his head, it was too big a glow for that, although only a splash of rose low down.

The next thought brought a gasp and a leap of
my pulse,—a volcano! "I think so," said my stolid
friend, to whose sailor mind land and sea were
phenomena only to be approached, passed over and
left behind. I turned and saw the dim loom of the
cliffs above me—so cold in the starlight, and the
thought of their beginnings took on a sense of
reality; there might be beginnings today.

Time after time I awoke and looked at the pink
blur, and once a great shooting star fell slantingly
into the very heart of the warm glow. It paled and
vanished in the dawn.

My first selfish thought was to stop scientific
work, drop everything and steam swiftly toward
the strange sight,—the exact effect of a fire alarm
on a school-boy. Then I found good and reasonable excuses. The thing itself was well worth veri-
fication, and the unexpected trip would interfere
not at all with my oceanographic work. We did
not wait even to hoist our flock of boats aboard,
but towed them ashore, dragged them to the top of
the beach and left them in the shade of the tent
flies.

The captain said that the direction of the light
was 256° of the compass circle, a bit of news which
was utterly unintelligible until correlated with the
more old-fashioned, visuable west-by-south-by-one-
half-south. Eagerly I laid down a ruler on the
Galápagos chart westward from Tower, and eighty
miles away it came to rest on Albemarle. It
touched the island far north of the five thousand-
foot southernmost peak which on the chart was

marked active, for reasons best known to the carto-
grapher.   It did not even intersect Narborough,
which, exactly one hundred years ago, had given
Morrell such a warm reception.

On our visit two years ago to these islands we
had hoped against hope for some hint of volcanic
activity, with the same chance of success that a
ramble through an old cobwebby attic might yield
some overlooked treasure of the past.   We calcu-
lated the half-dozen historic outbreaks to see if any
periodicity marked their occurrence, but all was
useless and we experienced nothing but the dead
cinders of a world's end.

With the present incentive to hope we cleared
Darwin Bay at ten o'clock and headed a little
south of west, steaming slowly over the calm
water.   James was on our port side, Bindloe to starboard, with Abingdon and Indefatigable dim in
the distance.   New soundings between Tower,
Bindloe, and James gave us data for our contour
map which was slowly taking form in the *Arcturus*
laboratory, and a deep haul brought up a catch of
strange deep-sea beings.   On we went, watching
clouds form and reform on the horizon, but never
certain of their origin.   Toward evening all the
islands, after flashing the sunset colors vividly
back and forth to one another, gradually withdrew
behind misty veils woven from their deeper valleys.
I put over a small net from the boom-walk and
skimmed the surface for an hour.   Then in the
dark-room I watched the wonderful glow from
the lantern fish which I caught—little eruptions of

body fires which flashed forth to their small world signals for food, for warning, and for a mate. Then I went on deck and saw the unmistakable glow of the fires of earth, and with a great wave of emotion I realized that my "World's End" had become World's Beginning.

All through the night we steamed at half speed toward Albemarle and every hour I went up on the bridge and focussed my high-power glasses ahead. From that same indefinite glow I had seen at Tower the eruption took form and size, and at last separate, gleaming lights could be distinguished.

Not satisfied with a single wonder, Nature sometimes takes us when we are immersed in the glory of some great sight and adds unexpectedly another, an auxiliary marvel for good measure, as a hint of the overflowing richness of the cosmic storehouse. Just before dawn, when three of us were watching silently with all our eyes, a mighty shooting star struck itself alight on the rim of our atmosphere, and in a blaze of white comet-light, fell silently and accurately into the center of the lava flow. After the identical happening of last evening, this appeared more than cosmic, it seemed intentional, and for a few moments I think the state of mind of all of us reverted to that of our distant forefathers, when signs and symbols and portents regulated all of life. When the possible combinations of the temporal and spatial arrival of a shooting star be considered, it was assuredly an astounding thing that two such mighty meteors should have taken this exact course.

The penalty or the advantage of human experience in various parts of the world is to stimulate similes of antithetical relationship. I have often laid at anchor off a gentle land, and before dawn watched the twinkling lights of villages and isolated farms go out one by one. Then the grey dawn picks out the larger hills and darker woodlands and finally the lesser objects. Farm houses materialize and from each goes up a twist of smoke, while that from the villages merges and floats slowly off as a unified cloud. Such was my introduction to the first volcanic eruption which I had ever seen actually come into being. Although the outward view recalled only such homely scenes, yet never was there absent from my mind its tremendous significance (Fig. 26).

Before the sun rose, the shifting light wrought another small magic. The lights had gone, the smoke had curled up peacefully for a half-hour, when I began to see that it did not twist and thread up as from breakfast fires beneath chimneys—it billowed and rolled. Then the houses dissolved before my eyes and became conical mounds or smoking ruins; what might have been fertile hill-slopes greyed to cinders, dotted with the upright skeletons of tortured or dead trees. Another casement of the magic window of memory opened, and I was lying before a mud-caked grating with the Verdun sentry, looking out upon an identical landscape: a few smoking ash heaps where once was Fleury, a cindery pile and a score of scarecrow trees in place of the peaceful beauty of Hadre-

court and Douamont. Before this spell broke I
went quietly down to breakfast.

When three miles off shore we sounded in a
mile and a quarter, and even when within a mile
of the coast we floated on a half mile of water.
In a fairly choppy sea three of us dropped into a
boat. As we approached the land we realized the
landing was going to be difficult. Heavy surf
dashed against the cliffs, or rushed madly over
half hidden reefs. Here and there were calm lagoons backed by mangroves, but always guarded
by deadly giant waves. Up and down the shore
we chugged, vainly looking for an opening. Twice
I almost made up my mind to jump and let myself
be washed ashore, but decided on a final reconnais-
sance to the northward toward Cape Marshall.
We passed deep caverns cut out of black lava and
once a natural bridge stretched across a gap from
which four sea-lions leaped down to have a look at
us. A golden grouper snapped at our shining brass
propeller, sharks cut outer edges with their fins
about the boat, and once a baby devilfish as large
as the door of my cabin playfully flipped one of his
wings and drenched us with spray.

As with all coasts, the capes and indentations
were visible only when we were opposite them and
hence we did not see a delightful little cove until
we had almost passed it. Turning sharply in, a
flock of noddy terns, pelicans, and some brown
boobies greeted us. As I was changing motor for
oars in shallow water a dozen big black groupers
rushed at us and bumped and bit at the oars and

keel. I have never seen such reckless voracity.
From appearances it would have been dangerous or
fatal for a man to have dived in at that moment.
Bill Merriam thrashed at the head of one with a
piece of canvas and it was torn out of his grasp
on the instant. I poked at them with the boat hook
and two seized it at once. Later a fisherman of our
party caught sixteen here,—a good two hundred
pounds,—on spoon or bare hooks, as fast as they
could be pulled in.

Before I left the *Arcturus,* I had very carefully
examined, with my number twelve stereos, the
whole of Albemarle in sight, and mapped out a
tentative route to one of the largest outpourings of
gas. At the outset I was handicapped by not being
able to indicate or speak of the two great mountains between which the eruption was in progress.
So I gave these nameless mountains the titles of
Mount Whiton and Mount Williams after the two
gentlemen without whose ship and generosity it is
probable that this volcanic outburst would never
have been recorded. The former is the most northern on Albemarle and its height is unknown.
Mount Williams is next, thirteen miles to the south
and over four thousand feet high, which is apparently somewhat less than the altitude of its neighbor. Most of the activity was along the slope connecting the two mountains, the actual glow from
the lava being visible in groups or lines rather high
up near the ridge. But at hundreds of places all
over the slopes were fumaroles, or cinder caves
from which poured forth greyish white gases. A

few openings over the summit of the ridge were high and large enough to merit the name of crater. Nowhere could molten lava actually be seen in the daytime.

I chose as our objective, a place of active eruption about half-way up the slope of Mt. Whiton. We found landing an easy matter in Eruption Cove, after we had picked our way over the broken reefs of coral and lava which guarded the entrance. Lacing on high, hob-nailed, moose-skin boots and carrying nothing but two empty snake bags and a single canteen, John Tee-Van and I set out this bright morning of Easter Sunday on the worst trip we have ever taken together. I have lost more blood from falls in a tramp over the high Himalayas, I have suffered much more from thirst in wild desert places of India and China, and have been more exhausted from lack of sleep during treks where there was no safe place to rest, but for sheer meanness and general uncomfortable travel this was the worst.

We started briskly with a last call to Bill in the boat to take us off in three or four hours. Our goal was unmistakable, for the underground powers had fired up and vast masses of billowing smoke were pouring forth.

The going at first was not bad. We had landed near the shore of a river of smooth, black lava about a mile wide, which had flowed seaward between banks of a rough, sharp pointed, apparently older flow. It was astonishingly like an actual stream or sea of water which, in the twinkling of an eye, had been transformed to a glassy jet substance. We passed over ossified ripples and swells and even curving waves with breaking tips so tissue-thin that light showed through them in a thousand places, and a slight blow of my hand broke off sheets several yards in extent, which clanged down into the hollows like steel falling upon steel. Sometimes we could pass dry shod like St. Peter over a wide stretch of calmer obsidian ocean, with here or there the fin of a shark or the head of a turtle protruding, or in a Jonahesque manner would chum familiarly with a mighty glass whale. Islands rose here and there, upon which perched great images of sea lizards and pterodactyls—all done in jet-black, molasses-like lava. It compelled steep up and down climbing, but was heavenly smooth.

Our fossil river grew smaller and soon petered out and we had to take to the real scoria; hellish rock froth which taxed our utmost strength. Imagine, if you can, a brobdignagian ploughed field and we two tiny ants essaying to cross it. But in place of soft and yielding earth, this was of razor-edged, needle-pointed clinker, sometimes steel-hard, again crumbling to a depth of yards. It was reddish brown and, unlike the obsidian river, had probably hardened slowly at the very surface. All the enclosed gases had thus had opportunity to escape, bubbling and blowing the cooling lava into thinnest crusts and skeleton rocks. The metal soil of this great ploughing was piled in pinnacles and mounds, brittle, sharp as knife-points and varying in size from a needle to a house.

At every step we crashed down through the mass as one might tread upon hill-sides of delicate glass, or we leaped unsuspectingly on a harder, steely stratum only to slip sideways or in turn bring down a lava slide upon legs and body. Often what appeared to be the softest turned out to be a solid boulder, and the consequent unexpected jar was more trying than a slide or slip.

We clung as much as possible to the smooth lava and by going somewhat out of our way were able to follow a narrow stream for a considerable distance. But sooner or later we had always to plunge into the red porous chaos. In ten minutes we were dripping and panting. The unclouded sun shone steadily down upon the sea of metal and soon there arose a reflected heat like the blast from a furnace. We headed steadily for the giant, outpouring cauldron well up on Mount Whiton's shoulders, reorienting our direction every time we climbed out of a furrow. Minutes passed, a half hour, and I realized that the simile of ants applied to our speed as well as to relative size. The coast seemed to recede with disheartening slowness, while the cauldron was as far off as ever. I decided to halt a few minutes to rest and found that even this was impossible. The heat from the lava when we stood still was unbearable, pouring up into our faces and scorching through the soles of our shoes. Even when we could occasionally find a smooth piece of lava, the stones were too hot to sit for a moment. I humbled myself and altered my objective to a lesser crater half as far away as the

large one, and after another half-hour's ghastly
toil I again surrendered and changed the angle of
our progress to the southwest, toward the nearest,
smallest fumarole out of which smoke and gas
came.

Every two hundred yards we stopped for a moment, standing and shifting from one foot to
another. I found that even a square foot of
shadowed rock yielded a welcome coolness to my
boots and feet, but we could not squat coolie-fash-
ion, for every breath of air ceased below a height
of three feet.

By the time I could distinguish the separate piles
of scoria around my small craters and the separate
jets of gas, the going got even worse, for now we
found our path intersected with ravines and cross
arroyos, the traversing of which was almost impossible. The last quarter mile I went ahead blindly,
and when I thought I must have reached the fumarole I found my way barred by a steep, unclimbable
cliff of crumbling lava, and far to the right a tiny
spurt of smoke. Disappointed, I turned to the left
and managed to surmount a thirty foot elevation
composed of scoria, breaking as easily as crackers
but of the hardness and sharpness of the steel resi-
due of factories. Fighting my way just ahead of
the avalanches of lava which I kicked down, I
came out on a flattened summit, and went on ten
yards farther. A glorious cool wind met me for
a moment, then died away and the sun's terrible
rays poured down, at the same time that twenty
fumaroles in all directions gave vent at once to

spouts of grey gas. Without knowing it I had climbed into the heart of the small, nearest crater which we had chosen. To escape the hot, terrible breaths of gas I stumbled forward to the eastward rim where four holes were evidently inactive. In a moment I realized my mistake and that I had entered the influence of some more awful invisible gas, perhaps carbon monoxide. The glaring sun became darkened for me and a frightful nausea forced me back to where the visible but less noxious fumes dominated. Added to this, the heat from below made the sun's influence seem almost benign. With my handkerchief over my nose and mouth I picked out several small pieces of lava covered with a whitish, crystallized exudate. Down one hole I could see a deep, rosy glow, but I could not stand the torture a moment longer, and half slid, half fell down the cruel, scrap-steel slope, and calling John, began our journey without a backward glance. We were too exhausted to do more than choose whatever way seemed least terrible. Now and then, from the summit of one of the dreadful furrows we could see the *Arcturus*—a tiny dot on the distant blue water, describing a five mile circle as she dragged a mile or more of deep-sea nets. Our drinking water was gone long before we returned and when we reached the shore we could hardly talk and were crumpled up with sudden cramps. I have had more than one strange Easter Sunday walk but never one like this.

Two yellow butterflies, one large fly and a few spiders near the shore comprised the fauna of this hell-like zone, while a single, daisy-flowered, aromatic shrub, and two half-burned cacti represented the outposts of plants or their forlorn hope.

As I lay on my back, half in the cool water, I heard the cry of a young pup seal, and in the cave of a tiny ravine just back of some mangroves I discovered the ideal nursery of the little chap. He hitched himself in, just out of arm's reach, as I approached. A hot breath of air struck on my neck and the quickened memory of the past five hours sent me quickly back to the coral lagoon, there to bathe until I left for the ship.

After eight glasses of water and a bottle of beer my aqueous equilibrium was restored and I studied the shore with the increased interest of intimate experience. I had acquired infinitely greater respect for the details of what met my eyes. I laughed when I thought how blandly I had chosen yonder crater far up on the slope as my goal, and then shifted to the comparatively tiny vent so near the shore, and which had proved large and dangerous enough to kill a hundred men in as many seconds, if they were to remain that length of time on the conical summit of the appalling gridiron.

I sought information from the best authority upon volcanos and at the outset was delighted to find the entire subject accredited in a most technical geology to a wholly heathen god of old—Vulcan. I felt that my own consummate ignorance of the subject was less reprehensible when I read, "For the present, volcanic hypotheses must work out their own destiny."

Years ago it was a terrible blow to have my theory shattered of a molten world, around which stretched a tissue skin of solid, cold rock on which we dwell, like mealy-bugs on an apple. With such a theory at one's beck it was so easy to picture the volcanic lava as simply flowing up through open pipes connected with this inner reservoir. But I have come to find an equal thrill in the more logical planetesimal idea, especially as it lessens in no way the possible number and extent of volcanic outbreaks in the future.

I like to think of the incentive to these miles-deep activities as residing at least in part in tidal stresses,—in the same pull of the moon as that which uncovers my tiny tide-pools. The great craters of Mounts Whiton and Williams are quite dead, choked apparently with solid plugs of lava flows, but the major part of northern Albemarle consists of the scoria, whose slow cooling, as I have already said, allowed much of the retained gas to escape, and left exposed the ploughed rock froth over which we had to toil. The porous character of this surface has precluded the blowing up of craters or ground in the present activity and has resulted in the intrusive type of irruption which I have described. The primary deep throat of lava flow must exist high up on the shoulder slopes of the two mountains, flowing thence beneath the surface, finding actual peep holes for the hot lava itself at scores of places, and sending forth the excess gases and steam through a thousand vents. There was a nexus of at least twenty-five of these, each a foot or more in diameter, at the miniature cinder cone which I reached.

The fascinating thing about the solid earth theory is the action and reaction of heat and pressure on rocks. If we penetrate the earth below the effect of seasonal changes the temperature increases about one degree in every sixty feet. Hence if the air at the surface is 70° Fahrenheit, at a depth of a mile it would be 158°. Carried to the center of the earth, this would reach the exceedingly warm temperature of about 350,000°! But the check to this explanation of molten lava is that, with the depth, pressure also increases from the earth's own gravity, and pressure is an absolute inhibitor of liquefaction. So as soon as we have gone deep enough to obtain the requisite 2000° to 3000° of heat necessary to melt rock, we automatically have a pressure which prevents it. But when old earth slips and shrinks, and surrounding hard rocks creep and give room to uncountable threads of liquid lava, and when the six mile zone of fracture beneath our feet somehow achieves direct touch with that, three or four times deeper, and the old mysterious tidal gravity gets in its work, up comes the lava to stir us mortals to our very souls.

So this is the story of my Galápagos volcano, which came to my consciousness with all the unexpectedness, and appealed to my enthusiasm and appreciation with all the power, of a single marvel, —at least that is what I thought as I steamed slowly northward in late afternoon. The sunset was directly behind it, and as the change was

FIG. 26.—BIRTH OF A VOLCANO.

Poisonous gas and steam ascending from slopes between Mount Whiton and Mount Williams.

Fig. 27.—*Arcturus* Staff Watching the Volcanic Eruption in the Galápagos.

wrought imperceptibly from pink and salmon sunset glow to the scarlet and white of the lava fires,
the cosmic splendor of the whole thing was overpowering. Whatever the theory of vulcanism, however learnedly we might discourse of lava and volcanos, light and the sun, the dominant thing was
that we had been brought close to the very beginning of things,—and this could not be written or
spoken, hardly thought indeed, but merely sensed
as one stood apart in a lonely corner of the deck.

But this Archipelago, when it had once opened
its heart to us who had learned to love it so, gave
lavishly, with measure overflowing. As when to
the volcano it had added the miracle of the shooting star and then duplicated this on the second
evening, so all the imagination of our company
combined could not have foretold what June the
fourteenth was to bring forth.

This was just nine weeks later, when we had returned from a trip clear to Panama to replenish
our stock of coal and fresh water. It was also
on a Sunday, when the *Arcturus* was again steaming along the shore of northeastern Albemarle.
The sun rose when we were exactly on the equator,
and the day broke clear and cool, with a strong
wind and current from the south. At seven o'clock
when we were all at breakfast, the wheezy, tin foghorn sounded from the bridge—a signal that something of interest was in sight. We all tumbled
up to see a great mass of steam pouring out apparently from the very sea beyond Cape Marshall.
For two days we had watched from a distance the

gas and smoke from the same craters and fumaroles which we discovered two months before. They hung in a dense, sickly cloud around the flanks of Mount Whiton, lower and yellower than the clean cloud wreaths which formed around the summit. During the two nights of observation of our former visit we had seen several new vents of lava light break out lower and lower on the slopes. And now the god or goddess of Great Desires had granted what must have been a powerful longing in our minds (I can answer for it in my constantly recurring thoughts) and after an interval of more than two months we were favored by being on the exact spot at the right hour; at last the living lava had reached the sea and we were the only witnesses in the world.

The Captain had first noticed the white ascending masses in the distance at six-thirty and thought it might possibly be spray thrown up over the rocky tip of Cape Marshall. Half an hour later, when he knew this could not be so, he trumpeted for us, and, bucking a strong head wind and a two-knot current, we steamed steadily ahead. I climbed to the rolling crow's-nest and in a wind which almost pinned my eyelids open or shut, I watched the puffing masses of white grow larger. For the first hour there was little change, and I utilized the advantage of my position as from an airplane, to watch the surface life of this deep blue water five miles off the coast of Albemarle.

Two or three large rays came flapping along— not the full-grown giant devilfish, but half-grown youngsters of the size only of an ordinary door and not a double barn-door. Now and then a sealion or two stood upright, half out of the water, gazing at us mildly, like stout little Balboas. The most wonderful sight was three huge *Mola,* or enormous sunfish. I had read, and seen pictures, of these massive monsters but this trio was the first in the flesh; and what flesh! They were devilfish stood on edge—oval masses, with tall dorsal fins, swimming upright, now and then veering enough to show the vast expanse of their vertical sides. I have seen replicas of their proportions in tiny half-inch larval fish which come sometimes in the surface trawls— unbelievably large around in proportion to their thickness (Plate III).

When Linnæus first saw one of these sunfish he seems to have exclaimed, *"Mola mollium!"* Millstone of millstones! And so ever afterwards, even until today, "Best Beloved," every ichthyologist repeats the exclamation *"Mola mola!"*

My recipe for making a *Mola* would be to take some enormous fish, of normal body outline, and chop it off just behind the short high dorsal and anal fins. Let these grow around the stump until they meet, and behold, a *Mola.*

Nearer and nearer came the volcanic outburst— ever more wonderful and awe-inspiring. We steamed as close as we dared, then turned and circled past again. This we did four times during the afternoon, then lay off-shore and made a last revolution after dark. At each perihelion we brought to bear our batteries of eyes, glasses, still and moving-picture cameras, and time after time, as the curtain of distance was raised, we felt we had front row seats at the most thrilling drama in the world. The current and the strong on-shore wind raised a sea which made launching a boat unthinkable—a bitter disappointment to me, who would have been glad to take greater chances than this for the opportunity of landing farther up the shore and approaching as near as possible (Fig. 27).

This was made doubly hard to resist when I looked along shore to the southward and there, only a few hundred yards away, saw the selfsame little mangrove-guarded cove where we had landed on Easter Sunday nine weeks before. The waves precluded repeating our visit, so we could only look with longing, and swing around for another broadside view of the new glorious outburst.

As we came closer, the amount and extent of up-pouring steam increased, actually as well as from the apparent change due to proximity. I noticed that there were irregular repetitions in its character. First a tremendous spurt of white, billowy steam would rush up into the air, tumbled and tossed landward by the strong wind; this would grey rather abruptly into a darker gas, then more steam, and so on. Seen dimly and at intervals through the steam, the high dark lava cliffs and levels showed for a considerable distance the same white incrustation of crystals which I had found around my fumarole on the inland slope.

Seizing a moment when the crow's-nest was comparatively steady, I swung my glasses along the line of juncture of steam and water and saw a curious red tinge upon a sloping rock. It was badly blurred, however, and I carefully cleaned my eye-pieces, and then saw that the red was fire and the blur was movement—and in the full light of the sun I watched an open artery of Mother Earth pouring into the sea—rock liquid as blood. The Galápagos were being born again.

Even at this early stage I fortunately realized that this wonderful dénouement of the April outbreak was only two hours old, and I watched the development and change of the various phases with a far more appreciative appraisement than would otherwise have been possible.

For example, when we first passed close along the whole front of eruption there were but six vents or rivers of lava, but before we left there were nine and a possible tenth. The outpouring steam and gas was at first about what might result from an enormous, shell-struck ammunition dump; the next time we circled near, it had quadrupled, and before dark it stretched out in a gradually enlarging cloud as far as the mid-slopes of Mount Whiton —a distance of at least eight miles.

Neither by day nor night was there any trace of live surface lava nearer than a mile to this coastal outbreak. A geologist would have calmly explained it as "An instructive example of an intrusive irruption changing into an extrusive eruption." But all a geologist permits himself when commenting on a volcanic eruption is the perfectly safe statement, "When molten rock is forced to the surface it gives rise to the most intense and impressive of all geological phenomena." My pity goes out to the student of earth and her rocks who has never yelled with sheer, unscientific, inarticulate enthusiasm at flowing lava, or been silent with awe at such a sight as confronted us.

When I recovered from the first great wave of realization of what good fortune had brought us, I perceived the astonishing details. The lava had crept slowly, week after week, down the slopes beneath the surface until it finally reached the end of the island flow. The amazing color of the whole was the most outstanding feature; the smoke, as I have said, was white and grey, the dead island jet-black, out of which spouted scarlet and white hot lava into water of unbelievable color. The sea around us and everywhere beyond the influence of this sudden eruption was a deep indigo blue, spattered and capped with white. When we first approached the lava streams, there stretched out into the blue water a narrow neck of clear, pure lumière green, enlarging at once into a shape which, from the crow's-nest, was exactly that of an old-fashioned powder-flask. For a time the Captain demurred about approaching this area, so perfectly did it resemble the green of extremely shallow water, but when within a hundred yards we could see that the surface agitation was the same as that of the blue water all around, and we knew its tint was due to some other cause. I have never seen two colors marked in liquid by so sharp a line. The normal temperature of the surface water off north-

east Albemarle that morning was 76°. When we
reached the boundary of green water most distant
from the eruption, I had Jay Pearson take temperatures as fast as he could pull up pailsful. We
were barely drifting at this time so that his records
covered a very short period of time and space.
When we were nearest the shore, although still
more than three hundred yards away, the green
water had risen from 80° to 99° Fahrenheit. At
this point we left the zone of influence and passed
into the blue water. At the moment when the line
of color was amidships the water under our stern
marked 99°, while that at the bows registered 78°,
a difference in less than two hundred feet of
twenty-one degrees.

I examined the heated water carefully but found
no sediment or suspended matter. A small tow
net drawn through it for fifteen minutes took only
a single blue copepod, a small *Coryphæna* or dolphin fish and a few tiny shrimps, all of which were
alive and well. Unless there was some inorganic
matter so fine that it showed no trace in a fresh or
a long-standing glass of water, the sharp color demarcation was due only to contained air or gas plus
increased temperature.

The streams of lava poured out of openings
several times their own diameter and soon formed
for themselves chutes of blackened, partly-cooled
lava. From time to time these nearly closed over
their streams so that three-fourths of a pipe was
formed, then as quickly the pipe would melt and
the great torrent stream out through the air unsup-

ported except by its own momentum. One spout
of lava cooled so quickly that twenty feet from its
appearance, great blobs of black appeared floating
on its surface—irregular, cinderous corpuscles tossing on this veritable vein of molten metal. Once I
saw a great lava river split into five separate
streams, which crawled down the hundred-foot
cliffs like the tentacles of some huge scarlet octopus.
These dripped down into the boiling green water,
while sulphurous fumes bubbled up in yellow froth.

There seemed a strange sort of irregular
sequence of force. First one stream would increase,
pulsating forth with greater violence, and immediately the sea would answer by catching great
quantities of the scarlet fluid and moulding them
instantly into gigantic, black bombs whose inner
gases would explode simultaneously, and shoot
forth a rain of half solid, half liquid boulder spray,
the jagged projectiles trailing comet-wise, fire, gas
and water in their wake. Then the steam from this
particular jet would billow up above all the others,
until a neighboring lava river in turn flooded its
banks.

We were close enough to see every detail, but
the fierce on-shore wind muffled every hiss and roar,
every bubble and crash, and we might have been
looking at the reproduction of some of the movies
we were taking. From time to time, a huge portion of cliff would seemingly rise a little, tremble,
and very slowly and gently topple forward, sending up a mountain of spray which alternately
crashed in great breakers against the living and

dead lava, and boiled and bubbled like some brobdignagian kettle. It was astonishing to see a swell
roll shoreward, curve up into a yellowish green
wave, shatter against the scarlet lava and instantly
rise and go floating off high in air toward the top
of the distant mountain. It was a battle, a cosmic
conflict among fire, water, earth and air such as
only astronomers might dream of or a maker of
worlds achieve.

Here at last was the very life blood of this Archipelago. Never would the black cliffs seem cold
and meaningless again, but always memory would
warm them and give them movement and color.
Their twisted strands, their broken, porous bombs
would seem to have cooled and exploded an instant
before; every gas-made tunnel might redden and
fill and pour at any moment.

I tried to estimate the speed of the lava and
chose a stream about twenty feet wide. As well as
I could judge at a distance of several hundred
yards the cliff at this point was about a hundred
feet high. I timed occasional black gobs of matter
floating down the trough and found that they traversed the entire drop in two seconds. Therefore
(as my old arithmetic used to say) *"if a stream of
lava flows a hundred feet in two seconds in one
hour it will,"* etc., etc., etc. My answer was that
the lava flowed thirty-four miles an hour. Here
was liquid lava in the open air and in a strong cooling breeze holding its two to three thousand degrees
of heat for a long distance, showing no blackening
before it was lost in the mass of steam and water.

What the temperature must have been underground to instigate such a cauldron is unthinkable.

The yellow froth near the shore seemed to indicate a considerable amount of sulphur and I knew by experience that those grey gases alternating with the steam were in part at least composed of hydrogen sulphide and carbon monoxide. I liked to think of the lava as causing real additions to our upper world—new volumes of hydrogen and carbon dioxide actually spread abroad in the atmosphere for the first time, and, before our eyes, rock substance changing from white, to scarlet, to pink, and to black, which since the beginning of the world had lain miles deep within its heart.

I have dwelt on the inorganic activity but, from the very first glimpse we had of the eruption, animal life was everywhere in evidence. Within two hours of its beginning, action and reaction had begun, direct and indirect effects on a host of creatures. A veritable black wave of fish passed us soon after we entered the green water—a school, or better a mob, of great tunnies, swimming close together with all their strength, panic written in every movement, headed for blue, cool water. Close to the gangway floated a great octopus, a yard long, half dead, his tentacles feebly moving, with waves of vivid color coming and going over his flabby body. A few small fish drifted by on their backs, and writhing, twisting sea-worms. In a small boat I could have learned much more of the effects of this rarest of rare phenomena.

Birds, to my surprise, were the dominating feature. While still a long distance away my glasses showed what I took to be shrapnel-like projectiles flung up and dropping down in the steam and lava. When closer, I saw that these were frigate-birds and shearwaters, not, to be sure, diving into the boiling water, but exceedingly close. Instead of the roar and rush of the unusual clouds of steam frightening away the seabirds, the sudden manna drew them in numbers, just as, when I use dynamite in collecting fish, the vultures of the sea gather at the first glimpse of a floating silver belly.

As best I could I made a census of the immediate eruption area and counted over two hundred and fifty stormy petrels, many in the dark phase, lacking the white rump, There were seventy-eight shearwaters of at least two species, thirty-six frigatebirds, ten brown boobies and three pelicans. Not only were they in the outer zone of green water but a dense flock was flying close in shore about the lava. All were attracted by floating fish or other organisms, and often I saw them actually become obscured by the steam and gases. Later two dead petrels and a shearwater floated past, so that some at least paid a price for their reckless search for food.

At the height of interest in this marvellous sight, but when we were at the aphelion of our circle, I watched the sea-birds through glasses and learned some facts new to me. The shearwaters not only flew in their usual erratic flight and snatched a morsel here and there from the surface, but they skimmed the surface with their beaks, ploughing it like skimmers. Besides this they flew actually into and through the high waves, working both feet and wings under water and often turning completely around before they emerged with tiny fish in their beaks. The wings flapped more rapidly under water and the feet paddled like mad. Every bird of the eight or ten near the *Arcturus* did this again and again, so it was in no sense an individual peculiarity. One shearwater was completely immersed for shorter or longer periods, seven times in nine minutes, and at the end the plumage seemed as dry as ever, and the flight was in no way heavy or impaired.

The greatest tragedy we saw was a full-grown sea-lion which suddenly leaped high, close to the shore. Five times he sprang, arching over eight to ten feet clear of the seething water and in blind agony headed straight for the scarlet delta of the lava. There was no final effort,—the last leap apparently carried him straight to death (Fig. 28).

At sunset we stood slowly in toward shore for a last look at the miracle which had been wrought for our benefit. I sat upon the very point of the bow and the sight which came to me from either hand might well have been from two different planets. To my left rose the long, sweeping slopes of Mount Williams, quiet in the sunlight, old, grey, dusty-looking lava alternating with masses of green cactus and bursera, while the shore was picked out with brilliant green mangroves. Clean, fleecy, unhurrying clouds drifted gently past the mountain's summit—Galápagos in her usual mood.

On the right, hell was let loose, a round worthy of Dante's lowest explorations—black, sinister crevasses, rushing steam, swirling ugly gases which swept on and on and finally joined the great noxious cloud which contaminated the clean mantle of Mount Whiton. In the foreground were scarlet, dripping lava and snarling bursts of gas-tortured bombs.

Dusk softened all this,—the gas vanished into the night and the nine lava streams became things of infinite beauty. The flying projectiles from the explosions were now seen as glowing red, not black. We turned and steamed toward James, and until ten o'clock that night, many miles away, the unforgettable fires burned over our stern. It was a wonderful farewell,—the very rocks of Galápagos alive.

Two things remain to be set down.

Twenty hours after we steamed away from Albemarle, our steering-gear, without a second's warning, broke down. Twenty hours earlier, with the violent on-shore wind and current, deep water up to the very splash of the lava,—and the good old wooden *Arcturus* would have contributed a new odor and a few flying sparks, and after that the steam and gas would have continued as usual, and the lava flowed uninterruptedly.

And now that I have had to reread all these words in hard type, I realize that I have given no more idea of the real happening than if I had attempted a description of the single peacock, the one opal, the solitary sunset which I had seen and you had not.

CHAPTER VI

OUR ISLANDS

BY RUTH ROSE

BILL MERRIAM was shouting from the foot of the gangway.

"Hurry up! the boat's ready! Come on!"

The anchor of the *Arcturus* had hardly splashed rustily into the placid waters of Gardner Bay when our flock of small boats splashed after it, and most of our land-hungry members eagerly sought for places in them. Soon the steep slope of white beach that fringes this side of Hood Island was dotted with exploring figures, scattering up and down the shore or vanishing into the thick scrub of the crater-side.

But even after weeks at sea, there were some of us who had decided to forego a shore expedition, at least until next day. During all our cruise I had listened to other people's fish stories, which is not meant in a derogatory sense; I had admired the shapes and colors of the fish caught by others, and had marvelled at the sizes of the ones that got away. But alone of the staff, I had never gone fishing,—not only on this expedition but in all my life. So under the kindly tutelage of Betty Trotter and

146

Fig. 28.—Lava from the Albemarle Volcano Pouring into the Sea, Killing Fish and Sea Lions.

The distinct line between the hot and cold water can be seen.

FIG. 29.—OSBORN ISLAND, BETWEEN GARDNER AND HOOD.
Part of a herd of tame sea-lions on the shore.

Bill, I had determined to sally forth today to catch a fish, not for science, possibly not even edible, but a fish caught merely for the sake of fishing. This much talked-of business had to be investigated and its thrills experienced.

So now, at the commanding tones that echoed from the ship's side, I hastily caught up a spoon that was not a spoon, a squid that was not a squid, and a large hunting-knife that was indisputably just that, and dashed to the little boat with its outboard motor.

In all the archipelago called Galápagos, there is no more beautiful spot than Gardner Bay. The wonderful shore-line of Hood Island, with a thousand fascinating coves, peninsulas, pinnacles and caves, shelters the smooth surface where rocky islets seem to float, like congealed drops flung off from the parent island when that was still a seething fountain of molten lava. The scars of the terrible searing floods that have poured over Hood, from summit to shore, are more nearly covered by vegetation than elsewhere in the group, and on this sunny April day the sea, sky and land seemed wonderfully new, a vivid picture-world that had not been created long enough to lose its delicious freshness.

The motor chugged us briskly to a sheltered cove, where Bill laid some deep plots against the lives and freedom of the crayfish in the shape of baited traps, and then we set off to the passage between Hood and Gardner Islands. At the moment which was mysteriously declared to be the

right one, Betty and I were bidden to throw over
the spoons and to let out what seemed to me like
several miles of line, and I was breathlessly embarked on my first fishing trip.

To begin by being breathless was a great mistake; I needed more breath than was available long
before we finished. There is such a thing as a
science in fishing, everyone asserts, so my very
brief experience must be misleading. There are
many strange things about the Galápagos, conditions that seem topsy-turvy to us, and the fishing
must share in this abnormality, for the sport of
angling in these waters seems to me to be mostly
an endurance contest, in which the fisherman sinks
from exhaustion or his boat sinks from the weight
of victims, not to his skill and cunning, but to the
mere fact that he can exert a few pounds more pull
on his end of the line than they can on theirs. The
real test of skill here would be to prevent the fish
from biting. My idea of fishing as a sport, solely
gained from one afternoon in Gardner Bay, is as
follows; you throw over a large, wicked hook, which
has an uncanny aptitude for turning and rending
you, and a shiny, curved piece of tin; you unreel
a lot of line, and wait thirty seconds. Your arm is
then jerked out of its socket, which you take as a
hint that a fish insists on fighting it out on this line
if it takes all winter. Your tutor in the gentle art
of angling then stops the motor, which saves you
from being dismembered. You start to pull in the
line, the fish registering violent disapproval and
arguing all the way. The line is extremely harsh,

and no one told you that gloves are worn when fishing; at frequent intervals you strike yourself on the chin with the large knobby piece of wood on which you coil each hard-won inch. At last with a rush, a great ugly head, with gaping jaws, pops out of water alongside, and standing up, to the imminent peril of the boat, you give one mighty heave and sit down suddenly, sometimes on the fish, sometimes with the fish in your lap. Naturally you emit a few piercing shrieks during all this, to the intense disgust of your masculine companion. In the process of recovering the hook and spoon, which have often been entirely swallowed, you acquire several wounds, and if I were writing a brochure containing Hints to Fishermen I should emphatically say, "Never, *never* get your fingers into the gills of a grouper." Then you gasp twice, throw out the line again and proceed as before. Now and then a fish, that by rights should be dead, slides stealthily along the bottom of the boat and deals you a tremendous smack with his tail. That usually incites the rest of the alleged corpses to imitation, and you feel like a Pilgrim Father running the gauntlet in a distinctly unfriendly Indian village.

Our catch consisted almost entirely of groupers, —big mottled fish whose voracity passes belief. After a while we found that it was not even necessary to troll for them; from the stationary boat the hook and spoon would be snapped up before more than a few feet of line were paid out.

Betty hooked one large Spanish mackerel, which put up a lively fight, and between us we also caught

five hieroglyphic fish, beautifully patterned with
cuneiform inscriptions that seem as though they
must be decipherable.

Now this business of catching large, resentful,
powerful fish as fast as the line can be thrown out
and pulled in, is excellent exercise and, for the first
few thousand fish, great fun. But after a while it
does pall upon one. As I dodged the assaults of
a hot-tempered grouper that was exhibiting every
sign of repugnance for the boat and our society, I
glanced above the level of the gunwale for the first
time in an hour or two. We were drifting in the
shadow of a cliff, and such a cliff! Sheer from the
water it rose, a black rampart to whose most impossible declivities clung little flowering plants.
At the very top, outlined on the cloudless sky, a
yellow-blossomed tree lifted thin arms, and the
clear whistle of a mockingbird drifted down.
Within oar's length a tiny pocket on the face of
the rock wall held a scanty nest, and the carnelian
eyes of the fork-tailed gull-mother watched us
calmly over her lava parapet. Just below, a low,
deep cave bored into the base of the cliff, and the
slow surge, creeping back, revealed glimpses of
rugged walls, softened and colored with the myriad
hues of bright sponges, starfish and anemones.
Sprawled motionless across the top of the arched
entrance, a giant black sea-lizard might have been
either a fairy-tale dragon guarding his den, or the
sculptured device of an artistic Prospero.

My lagging interest in angling died altogether
and I looked about to orient myself. We were

between Gardner Island, the largest of Hood's satellites, and this cliff, which was the face of a jutting point on an islet between Gardner and Hood. Its height was as out of proportion to the diameter of the island as that of a skyscraper. As we turned back toward the *Arcturus* we crossed a shoal that projects from the south side of Gardner, where each big roller, as it piled leisurely against the obstruction, showed in its curling green arch a dozen groupers apparently enjoying the sport of surf-riding.

As soon as we climbed aboard the ship, I went to the chartroom to find out the name of Islet-South-Of-Gardner. We must have been the first visitors to take an interest in it, for on none of our maps or charts was it considered worthy of more than anonymous delineation.

Everyone knows the fascination of the miniature; witness the steady market for ship models and Japanese toy gardens. It is a kindred feeling that makes islands more attractive than continents, and the smaller the island, the greater its charm. Next day we were in the diving-boat in the lee of Gardner, and the Unknown Isle loomed across the intervening strait, looking more and more mysteriously inviting with every passing hour. By the time my turn came to don the mediæval-looking helmet and climb slowly down into the misty blue-green world of water, this apparently unimportant bit of land had become the most desirable spot in all the Galápagos. But something always happened to prevent a visit; no small boat was available, or there was something very pressing to be done, and after a while I began to go about murmuring, *"Je n'ai jamais vu Carcassonne!"*

It is a great thing to have authority on your side, so when the Director took an interest in Islet-South-Of-Gardner I finally reached it. One morning he and Betty and I were ferried over; we landed on a little lava step to which it was just possible to jump from the stern of the small boat. This was on the opposite side of the island from the cliff, and seemed the only feasible landing place, for to the left the shores were too precipitous, and to the right a long arm of boulders was partly covered by a turmoil of surf. No one expects to land on a Galápagos dryshod; it is counted a lucky day when an effected landing leaves you dry from the waist up. So we dripped moistly up along a series of zigzag shelves in the rock, until we stood on a level bit of soil. The whole islet, seen from this point, seemed to slope gently from the northern to the southern side,—from the high point of the cliff down to the boulder beach that buried itself in the sea. Here among big water-smoothed stones were other lumps,—some dark as the lava rocks, others yellow-brown, depending on whether the sea-lions had had time to dry since heaving themselves out of the surf.

From my short acquaintance with the race, I feel justified in generalizing to the extent of stating that sea-lions are nice people. From the chunky unweaned babies that can be tucked under the arm and lugged about, somewhat cumbersome but very lovable, to the old bulls with bristly moustaches, who pretend to be dangerous but turn sheepish when outfaced, they are all amusing, and some of the most delightful hours in the archipelago I have spent in their society. So now I naturally swerved toward the bulky bodies sprawled in the sun, all sleeping as blissfully as though lava boulders made the softest bed in the world.

The first group that I approached was of mother and child,—the latter enjoying peaceful dreams in a small tide puddle, while the guardian parent, a few feet away, lay dozing with her chin propped on a keen-edged stone. I walked up behind them, sat down a yard away, and remarked gently, "Good-morning."

There was no reaction to this, except that the skin on the mother's neck twitched, as though my voice were a ticklish sort of fly. I repeated my greeting somewhat louder. The otter-like head lifted from the rocky pillow and slowly swung in my direction, and not until both eyes were brought to bear upon me did the full horror of the situation dawn upon her. With a mighty snort of amazement, she sat as bolt upright as a sea-lion can sit, and braced on right-angled wrists she stared transfixed. I stared back, for one of her eyes was a repulsive, sightless mass of mucus, the result of a wound, I thought at first. The baby had not moved, so I reached out and patted his little rump. He rolled over nearly on his back, waving a languid flipper, but when he saw me, he went into reverse and lumbered hastily to his mother's side.

After spending five minutes in deliberate inspection, she decided that I was too strange a creature to be a desirable associate and withdrew, her offspring shuffling laboriously behind her, to a more distant spot, where they sank down and instantly went to sleep again.

My next attempt to be accepted in sea-lion circles met with more success. Further down the beach there was a group of youngsters, with one adult female apparently in charge; as a sea-lion has only one pup at a birth, I could only suppose that this was a sort of *crèche,* where other mothers left their children under the watchful eye of a good-natured neighbor while they went out to do the fish-marketing. Here I was received, if not as an equal, at least with more toleration than before, and as children are notoriously less suspicious than their elders, these fat sleek pups believed in my good intentions. The nurse or mothers' helper or whatever she was, showed some uneasiness at my too-familiar approach and at last made slowly for the water with her brood in tow, but when I followed and stepped into the shallow tide-pool in the rocks, which was evidently a favorite play-ground, the young ones floundered around me, lifting their small whiskered muzzles to peer curiously into my face as I crouched, and unmercifully tickling my bare feet and ankles as they dived to investigate my submerged portions.

As I looked from one to another of the doggish faces, I realized that every one of these pups had something the matter with its eyes. A sea-lion's eyes

out of water have a dim, near-sighted look, and as
they dry in the air there is often some whitish matter about the corners, but each of these babies had,
to a lesser degree, the same affliction that had
made that first mother partially sightless. I went
ashore to investigate, and of all the sea-lions on that
little beach there was hardly one, old or young, that
was without this disease. None of us had ever seen
this before, either on our previous expedition of
1923, or during the present one. Here was something like a leper colony, composed almost entirely
of the diseased animals, although this apparent
segregation was probably more accidental than intentional, a voluntary rather than a compelled os-
tracism. The most pitiful sight was a small pup
that was quite blind. He lay at some distance from
any others, seemingly as well-nourished as any of
the healthier babies, so the law of Sparta is evidently not in force among sea-lions. He was more
frightened than any of his fellows when I approached, and before I really touched him, he
began to scramble frantically away, crashing
among the stones so recklessly that I hastily retreated, lest he should hurt himself or stray too
far away from the spot where his returning mother
would expect to find him.

He was being taken care of now, but I wondered how he would fare when next year there was
another pup, a new arrival that would claim all
the mother's attention. However, his future was
settled out of hand when Dr. Cady heard of this
island isolation ward. Next day the blind pup was

secured for examination, and the disease diagnosed as conjunctivitis. How this was ever contracted by Galápagos sea-lions no one has explained.

In fiction certain conventions are always observed by castaways upon a desert island. The first thing they do is to make a circuit of its shores, attaching names to various bits of geography as they go. Wishing to do the thing according to the best traditions, Betty and I set off to explore the coast-line. The Director was just visible in a tangle of scrub half-way up the cliff, and from his immobility we knew he was watching some creature, probably a nesting bird. At such moments in the life of a naturalist, the advent of spectators is seldom hailed with enthusiasm, so we discreetly left him to his observations. A few steps beyond the boulder beach brought us to a steep rock slide, worn glassy smooth by the sea-lions that had glissaded down its slope. Descent was easy, merely an imitation of the sea-lions. We landed at the foot of a cliff, where a big black lizard was spread-eagled against the lava, looking like a skin pegged out to dry in the sun. Our somewhat hilarious arrival disturbed him and he straddled up the face of the cliff, clinging to invisible projections with strong curved claws. Just to prove that we clawless beings were not wholly incapable of acrobatics, we swarmed after him and caught him by his thick, serrated tail. Once captured, he hung limply in our hands, resigned to fate, and even when replaced on the cliff, he remained quiescent as though incredulous of his good fortune.

The Amblyrhynchus of Hood and its surrounding islets is not of somber, unrelieved black, as are most of these marine lizards of the archipelago, but is irregularly streaked with dull red in varying quantities. The simile that occurred to us at the time was of a neglected rusty suit of black armor; a few weeks later and we would have said that the lizard repeated the tones of a volcanic eruption, seen in full sunlight—the old, cold lava for a background, trickled over by streams of molten lava.

Along the shelving shore ran a narrow path, a ledge cut in the coarse red rubble. It looked like a mountain-side sheep trail, and on one of the larger islands we would have supposed it to be a goat thoroughfare. But when we had to climb over boulders that jutted across the foot-way, since there was scarcely two feet of space beneath them, it was easy to see that no creatures of goat's stature ever wore this track. Below, the creaming surf whipped round a thousand little crags and promontories, where pompous pelicans watched for delicacies and took off clumsily in pursuit of them. Big scarlet crabs spangled the black rocks, and vermilion-throated sand lizards scampered after insects. The path dipped steeply and stopped at a pebbly beach, shut in all round by high rock walls. A tiny pool, left by the tide that had crept away through some invisible crevice, was occupied by a half-grown sea-lion, and a few yellow-tailed fish that were too small to interest him.

By this time we had worked round toward the lofty side of the islet, so that the land side of the

beach was faced by a sheer high wall. And in this
wall was a low-arched opening, as black within as
without. Not daring to hope that it would be anything more than the merest recess worn by the
waves, we entered. To our delight (for what could
be more satisfactory than an unexplored cave on
a desert island) it was a narrow passage that turned
sharply from the entrance and seemed to continue
for some distance. At first we could stand erect
and walk over a thick strewing of round pebbles,
but presently in growing darkness the roof came
down so low that we took to all fours and crept
along a tunnel, dimly lighted now and then
through fissures in the rock wall, opening at the
level of the sea outside. A flickering greenish light,
cast by the reflection of the sun on shallow water,
made our crepuscular worming even more eerie
than it would have seemed in total darkness. The
silken sound of the wavelets slipping over stones
inspired us simultaneously with the thought of the
tide, and in involuntary whispers we discovered
that we did not know whether it was rising or falling. Somewhat reassured by the recollection that
this was no Bay of Fundy or Mont St. Michel, and
that we should probably have sufficient warning to
escape from these confined quarters, we crawled
on. Presently the roof sloped up again, so that
we could thankfully rise from bruised knees. It
was very dark now, but stretching up and to both
sides, no walls or roof could be reached. We turned
round a slight angle and came suddenly into a
large chamber, where at the further end a mys-

terious beam of light fell across a snow-white dais.

Rider Haggard could have done no better. As we stood stock-still hardly venturing to breathe, the consciousness of things moving quietly all about us was conveyed by more than the sense of actual hearing. Soft sighs, the rustle of a displaced pebble, a queer sibilant little sound between a breath and a hiss, peopled the gloom and sent tingles up and down our spines. Wild thoughts raced through our minds,—gnomes, mermaids, strange island folk of unhuman ancestry, or something too weird to imagine with even so much definiteness. Then a warm, wet nose sniffed experimentally around our ankles, and almost before we had time to realize that our cave trolls were sea-lions, and the white throne a wave-washed pile of pebbles and coral, there was a clatter and tinkle of stones, like faint cymbals and timbrels, and into the beam of light across the pale divan came the biggest sea-lion I ever saw. The circumstances and surroundings conspired to make him even larger than he would have seemed in daylight, I suppose, but he was assuredly the great-grandfather and the king of all his kind.

He advanced to the exact center of the spotlight and posed there. It seemed as though some one ought to cry "Oyez! Oyez!" but the only sounds were the subdued sighs under our feet and further back in unseen recesses the sibilant noises made by suckling pups.

The chamber was partially divided by a low wall running down the middle and we leaned on this barrier, occasionally clutching each other to express our utter satisfaction. The light came from a sort of chimney-hole high up in the seaward wall, and on a shelf beneath the aperture lay a sea-lion, for all the world like an electrician in charge of a theater spotlight perched in his little box halfway up the proscenium arch.

The king seemed quite unaware of us until we slowly approached the throne; perhaps we did not observe the ceremonial proper to such a progress. Suddenly the patriarch emitted a terrific snuffling bellow and hurled himself straight at us, to the accompaniment of an avalanche of stones. Outside we should have known that he was harmless, but this stage setting was too much for our nerves. With muffled howls we threw ourselves prostrate on top of the dividing wall, hastily elevating our feet, and the monarch of the den thundered past. The complete humility of our concerted salaam should have placated him. We heard the echoes of his indignant wheezes dying away down the tunnel, and with a flop the royal electrician deserted his now useless post and shuffled in pursuit. This was the beginning of a general exodus. The harem, or courtiers, did not seem afraid of us; but it was perhaps court etiquette to follow the royal suit. In five minutes we were in sole possession of the audience-chamber, except that from corners too deep to be penetrated by the dim light still came the sound of happily-nourished infants.

We introduced the note of a Christmas pantomime into this equatorial fairy-tale by leaving through the chimney, which was a scramble and a tight squeeze before we emerged on a rock platform halfway up a cliff and blinked in the blaze of sun. Feeling the combined sensations of Ali Baba, Tom Sawyer and the first explorer of cave-dwellings, the thought that made our enjoyment the more keen was that in all probability our feet were the first human ones to explore this island. There have been plenty of visitors to Hood, and scientists have collected from Gardner as well, but so far as we know, no one ever troubled to investigate these smaller islands of lava. Pirates and whalers, of pre-scientific days, might have landed on them, had they wished, but those strictly utilitarian gentlemen would have had no reason for doing so, though the sea-lion cave would have been an ideal hiding-place for treasure.

Continuing the circuit, we paused in some tide-pools, where we earnestly attempted to capture the wariest small fish I ever saw. A blue-footed booby watched us superciliously from an over-hanging ledge, with an air of I-could-an-if-I-would. Presently our fruitless efforts were interrupted by shouts, and the Director rushed to join us, all agog with the tale of a marvellous cave that *he* had discovered. We gave an imitation of the booby's expression, and explained carefully to him just what he had missed by not being with *us* when the cave was really discovered, instead of merely following in the footsteps of the pioneers, and finding that civilization had driven out the aborigines.

From a jutting headland we looked down to the

sea over a straight drop, a wall which gradually
rose to the eminence of the cliff which had first attracted me to the island. Prospero's cave was not
visible from this point. The air was sweet with the
odor of a shrub with racemes of greenish-white
flowers, and three or four bees hummed over the
lures. These bees and these inconspicuous plants
were the Galápagos manifestation of tropical luxuriance, the best that the islands could produce.
There were many small, drab moths, and some of
the low, pale-barked trees were almost leafless from
the depredations of little green measuring worms,
presumably the larval form of the moths. A small-
billed finch was patiently stuffing her clamoring
full-grown attendant youngster with as many of
these worms as her careful search disclosed.

Entomology in the Galápagos must be pursued
by painstaking examination of every leaf and twig,
hole and corner, crack and crevice. Turning over
large stones is one way of collecting, and we applied ourselves to this grubby method, on the steep
hillside among thorny scrub and cactus. Two large
reddish centipedes were our first reward, a dubious
delight to the non-naturalist, but most welcome to
our collecting bottles, as the only specimens of an
equal size that we had so far acquired had been
by the fragmentary method of taking them from
the stomachs of dissected hawks. The first scut-
tling rush of a tiny gecko was hailed with shouts,
for these little lizards are rather rare on the islands,
and we were doubly interested in seeing what differences there might be between those from this

lesser land and their brothers on Hood. Diligent
search, interspersed with bursts of speed, gave us
four geckos, and we saw three others that were too
agile for our combined efforts. A hawk soared
overhead, perhaps watching our pursuit of such
edible morsels as centipedes and lizards.

In certain directions the Galápagos is a narrow
field of research. For instance, if you have seen a
hawk there, you can rest on the assurance that you
have now seen every species of hawk to be found
in the archipelago, and the same is true of an owl.
Those bees whose busy wings buzzed companion-
ably about us are the only representatives of their
family on the islands. But on the other hand, of
the black finches that are native here there are at
least fourteen species (Fig. 29).

Several kinds of beetles, a nest of flying ants and
one of termites, and a white, thread-like centipede
were disposed in vials before the blare of the ship's
whistle warned us to be ready to perform those
athletic feats necessary to embarkation in the small
boat. Returning with what we felt to be an almost
complete collection of the flora and fauna of our
nameless islet, we christened it Osborn Island for
Professor Henry Fairfield Osborn, and fell to
sorting and preserving our specimens for future
study. The Director combed Osborn Island with
especial thoroughness for the total bird census, and,
to add to the scientific value of this humble chapter, I have prevailed upon him to condense some of
his observations into Appendix A at the end of this
volume.

But now we sighed for smaller worlds to con-
quer, for from the rocky shelf of Osborn Island
we had had an enticing view of an islet of even less
acreage.   Attraction and size seemed to be in in-
verse ratio; Hood would certainly rank as a small
island; it is quite invisible on most world maps, but
when anchored off its shores it loomed large, with
the majesty of a continent, since the eye could not
compass it at a glance.   Probably I ought to be
ashamed to admit that I never set foot on Hood
during all the time that we lay in Gardner Bay.
Gardner Island, a sort of New Zealand to Hood's
Australia, held something more of island lure; the
key to that fascination must be the hope that one
may more fully possess it through a fuller knowledge, which is, after all, the only real kind of possession.   Osborn Island had drawn us with the
promise of imparted secrets, and now there was a
smaller scrap, of infinite possibilities, with a delightful definiteness of outline that assured us of
complete results in exploring.

At dawn next morning the patient doctor was
routed out; he and one of the motor-boats understood each other, and he was accordingly elected to
the position of the most popular ferryman.   He dis-
played the spirit of a true Christian about it, and
while he nourished the engine with gasoline, Lin,
Betty and I raided the galley for our own sustenance.   The baker, a fat man with a grim face and
a kind heart, thrust a slab of coffee cake into our
hands, adding a festive touch to the humble bread
and butter to which we helped ourselves.   A can-

teen of the rather highly flavored water which was the best the condensers could distill completed our modest ideas of a picnic, and even before the early breakfast hour the fussy sputter of the outboard motor was profaning the crystal stillness of the bay. A grey sky slowly burned to blue at the zenith, while all round the horizon streaks of color brightened and faded. The smooth grey water looked so solid that the boat's prow seemed to carve a way through a leaden sheet that fell back in long wrinkles.

Our splashy landing on a submerged ledge did not even wake a half-grown sea-lion from his beauty sleep and with a sympathy in our hearts born of many reluctant risings of our own, we left him in peace. The receding gasps of the motor were swallowed up in the vast quiet, and as the first cool sunlight touched our Terra Incognita, we sat on a patch of scanty grass to breakfast, and tossed crumbs and crusts to a perky mockingbird and a big scarlet crab that hopped and sidled round us.

Near at hand were tide-pools where tiny bright fishes hurried about and maroon-and-pink anemones closed flabbily over my intruding fingers. Following the northern shore, we climbed along a rapidly rising series of huge steps roughly formed of crumbled rocks and reddish, friable earth. Over the edge of the cliffs we could catch glimpses of cosy homes tucked away in miniature caves or on hollowed ledges, where fork-tailed gulls and noddy terns were bringing up their families, and cocking

unconcerned heads at the apparition of our interested faces appearing from above, usually up-
side down. A beautiful little red-footed dove fluttered from under our feet, and added to her plump,
partridge-like appearance by simulating a broken
wing in an attempt to lure us from the two white
eggs lying under a boulder on a heap of twigs.

Presently we came to such a barrier of thorny
scrub that we went inland, still aiming for the highest point of Our Island. Its center was a cup-
like depression sloping toward the sea, and here
an incongruous memory smote me. The evenly
spaced, low, gnarled trees, the seeded grasses growing long and rank beneath, the rocks lying in tumbled lines here and there, strangely resembled an
abandoned New England orchard with crumbled
stone-walls and once-cultivated air. There was
even a sort of pit which needed only a sidewise
glance to be a cellar, the forlorn remnant of a home,
such as one finds on a country back-road, or comes
upon in a short-cut across young timber-land.
Surely an unexpected comparison, for of all places
in the world the Galápagos and New England are
the least alike.

The sunlight was no longer cool, nor were we, as
we struggled up the steep inner side of the cup,
cut off from any breeze and lacking anything that
could be called shade. A final scramble and we
emerged on the pinnacle rim to a panorama that
made us gasp. Far off was the misty loom that we
knew for Chatham Island, twenty-five miles away,
while close at hand were the shores of Hood with

YOUNG FISH TAKEN AT THE SURFACE IN MID-OCEAN

FIG. A. Young of the Giant Sun-fish, *Mola mola* (Linné)
"   B. Young of an unknown Soldier-fish, *Holocentrus*
"   C. Young of an unknown Pomfret, *Taractes*
(Actual length of all three fish, one-half to three-fourths of an inch)

PLATE III

FIG. A

FIG. B

FIG. C

a lacy edge of surf. Gardner and Osborn, to the north and west, were little dark heaps speckled with faint green, and a toy *Arcturus* lay between them. Rounding the furthest point of Hood was a moving dot,—the launch returning from a visit to the albatross rookery. And everywhere stretched the empty miles of blue plain, the summer sea, ruffled by a lusty trade-wind that fanned our hot faces with sweet air.

The sweeping view held us so long that we did not for some minutes see what lay below. At our feet was a sheer drop of a hundred feet or more; a long jetty, a narrow rampart of rock, perhaps thirty feet high, projected from the island for some distance, and turned at a neat accurate right-angle to parallel the line of the cliff where we stood, so that we looked down upon a perfectly protected harbor, enclosed on three sides. It hardly seemed possible that man had had no hand in the shaping of this precise alignment, that might have been a miniature of some such famous port as Alexandria, where the populace sauntered on the mole and watched ships come and go. In this case the gossipping crowd was composed of boobies, jostling each other along the narrow wall-top or standing stiffly like sentinels silhouetted against the sea.

They had no ships to watch, but there was activity enough, of a kind that was as visible to them as it was to us from our loftier perch. At the surface floated an enormous school of the beautiful white-striped angelfish, *Holocanthus passer,* their black bodies, splashed with orange, red and purple, plainly seen as they lay on their sides in the fashion peculiar to this species. We had seen them before by the half-dozen close about the small boats along shore, but there were hundreds in this group, drifting like bright petals in the clear deep water of the sheltered haven. Beyond the natural breakwater a round dark object seemed to be a rock, just awash, until two flippers shot out to propel it forward and turn it into a huge turtle. A school of large carangids drove swiftly past like a cloud shadow, and the sharp fins of three sharks cruised aimlessly to and fro above the long, submerged bodies. Across the strait between Our Island and Hood came thirteen great bat shapes,—rays swimming with graceful undulations of their wing-like fins, and holding their places in accurate battle-squadron formation.

From our height the water seemed a medium scarcely thicker than air. Now and then a booby launched himself from his observation post, soared rapidly and fell like a stone in a breath-taking dive. A slight splash, the bird disappeared while you could count three, and then popped out like a seed squeezed from an orange. A few steps further on we reached the best place of all. Behind the jetty stood a rock pinnacle, broad at the base, flattened on both sides, with razor edges, and tapering to a point that was almost as high as our cliff. It was a gigantic arrow-head, standing in the surf, and connected with the island by a tiny causeway far down at its foot. The waves boiled in a narrow tunnel they had worn completely through the base of this stone triangle, and on its very apex perched a lonely gull.

We sat on the brink of the precipice and with heads tilted far back, watched a frigatebird soaring overhead. There was something hypnotic in the unceasing song of the wind, the abyss below, and the vast blue vault above, empty save for a pair of outstretched wings that rocked lazily round and round a wide circle. At long intervals those wings flapped twice, then stiffened and held motionless, while the bird, confidently cradled on the rushing air, swung in its chosen orbit and watched its world. There was a dizzy moment when we too seemed to wheel in a great void, when, gasping, we clutched the solid rocks beneath, and brought our eyes, and so our bodies, back to the reality of finite earth and ocean, surf and sand.

Sometimes, in the confusion of cities, in the midst of the dirt and noise and countless irritations that make civilization seem a deplorable blunder, it is good to remember that a frigatebird is winging over that little secret harbor which, it may be, was never seen by any other human eyes than ours.

CHAPTER VII

WITH THE SHARKS OF NARBOROUGH

I ANCHORED the glass-bottomed diving boat as close to the cliffs of northern Narborough as I dared, in a cove where the water was so deep that the swells remained unbroken until shattered against the lava itself. The rocks at this point showed very clearly their division into successive lava flows, some like frozen, black molasses candy six feet thick, alternating with thinner strata in the shape of huge bricks. The topmost layer was the same old ploughed field of cinder crags and snags with which we were so familiar on Albemarle. This is probably the eruption of one hundred years ago of which Morrell wrote so vividly.[1]

This, my seventeenth descent, took me into a submarine world as strange and as unlike that of Tagus Cove (which we could still see in the distance from the ship), as that differed from Tower. If they were jungles and deserts this was a wheatfield. Swallowing as I went, I climbed down and down and stood, at last, on a gigantic rounded boulder, thirty feet below the surface. This roundness spelled a distinct difference between this and

[1] Galápagos: World's End, pp. 401-405.

other shores of the Galápagos. The surf had
pounded and rolled the rocks on this unprotected
coast until they had become huge pebbles. This
explained the absence of tide-pools along the shore
—the water simply filtering away as soon as the
tide level went down.

The dominant note of this under-water scene in
this marvellous island eddy was the sea-weed.
Great fields of it extended to the limit of vision,
with bare or sponge-covered boulders between.
Sargassum with small berries, grew on long, slender
fronds, two or three feet in length, which gave completely to every surge, more so than any land
growth to the wind. While I have dived where
steady currents hold in one direction day and night,
yet by the very force of circumstances, my puny
efforts are usually confined to the surge-affected
shore. Like a tide which changes every twelve seconds instead of every twelve hours, the whole
underworld swayed outward and then, with infinite grace, inward again. All of the innumerable
strands of greenish olive bent and flattened away
from me, and then, with the slow movement at-
tained only rarely by such growths as weeping wil-
lows, rolled toward and wrapped around me, reaching out toward the steep ascent marking the
beginning of that upper world which seemed so
little a part of my life at a moment like this. As
the grass shifted and vibrated, many weird little
inhabitants were disclosed for a moment, and then
scuttled back to shelter—wrasse never seen before
or since, twisting worms, crabs and snails, all iden-

tical in color with the weed. The numbers and
size of the fish outside the weed were remarkable,
almost every species being represented by larger
individuals than elsewhere, due perhaps to the unusual abundance of food on these current-served
shores. My old friends *Xesurus,* the yellow-tailed
cows, were grazing in schools of two to three hundred, shadowing slowly about the corners of boulders.

I was half way up a steep slope, and by twisting
the boat around with me I succeeded in reaching
the summit, where I could look down upon a
sinister valley, narrow and dark and deep, with the
opposite ridge covered with the same long, waving
weed. As I stretched full length upon a mat of the
sargassum, a gang—they were too ugly and dangerous looking to call school—of giant groupers parted
the fronds and drifted through toward me, all dark,
in tone with the olives and browns. They mouched
along, their ugly jaws chewing eternally on the cud
of life, when suddenly, without the slightest warning, there came a distinct glow and next to the last
grouper came one of the goldens. To their evident
opinion there was no difference; he impatiently
nudged a neighbor and in turn was pushed aside
by the fish following him. The most careful dissection on our part shows absolutely no physical
difference and yet, instead of being clad in mottled
olive green of the dullest, darkest shade, he is solid
gold from mouth to tail. The weed was appreciably illumined when he passed through it. One
strange thing has been that, rare as these golden

groupers are, both two years ago and during the present trip, it is only these gorgeously colored individuals which attack the propeller of our little outboard motors. Whether the color of the glistening brass attracts this shining caste more than it does the other, duller grouper persons, I have no idea.

A few minutes later a shadowy school, a second lot, of even larger groupers swept past in the blue distance with another golden brother with them. He is all the more wonderful because there are no intermediates—one has either regal golden blood or mottled brown polloi caste. Here is materialized the mental effect which creates in fairy tales the one most beautiful creature or hero or princess among a host of dull or ugly ones.

Once again a huge sea-lion gave me a start. As I stood watching a mist of grazing *Xesurus* I felt a sudden water pressure against my back and legs, and turned in time to see a monstrous black shape bank and veer away, having rushed down in a lightning sweep within a foot of me. His eyes were no longer the dull, soft, deer-like, half-seeing organs with which he gazed at me on land, but bright and clear and keen; the long cheek whiskers stood out white and bristling, the mouth partly opened as he turned and the dog teeth gleamed wickedly. As my eye caught his form I leaped involuntarily toward the ladder, forgetting that I was in a land where mighty acrobatics could be achieved with a mere push. I landed on a boulder at a height of about four rungs up, and some eight feet beyond the ladder—a standing high and distance jump which broke the world's record in the upper air by feet. The strangest thing about it was that whenever I did such a thing as this, I accomplished it slowly. I took off with deliberation in spite of my strongest effort, I went through the water with conscious lapse of time, and I landed as in a slow motion picture.

The instant I leaped I realized my mistake and watched the wonderful form as it swung up from me. It turned just below the surface and again shot down. I think a considerable percentage of these manœuvers was pure side, executed for the benefit of a smaller, probably a lady sea-lion, who hung between earth and air a short distance away, and watched. The big male—he was certainly over seven feet long—began his second rush at an acute angle, heading for the bottom some distance away. Turning like a meteor the moment his head touched the waving seaweed, he again cleared me by inches. I could not help but flinch, not from a fear of being bitten but from a disbelief that such a great body could possibly stop its impetus and not smash into me. As he passed, I stretched out a hand and felt the smooth, hard body brush against my fingers. This was apparently a surprise to the animal, who, in alarm, inserted an extra curve into his simple parabola, and in the effort gasped out a mouthful of bubbles. This time he shot to the surface and half out, followed by his admirer, while the string of bubbles ascended slowly—coalescing as it went into larger and fewer spheres—like the puff of smoke from an airplane engine, or the blossoming of white shrapnel against a blue sky. In each bubble I could see a distorted reflection of myself, my helmet and all my surroundings.

A glance around showed that every fish had vanished, and not until two or three minutes had passed did they begin slowly coming into view. The sea-lions are the masters of these waters, and I was surprised to see even a great turtle slide hastily out of the way when one came too near.   Sharks always disappeared with the fish.

Even if the fish had not returned I could have watched the movement of the seaweed for hours, it was so unlike the movement of wheat or grass. The whole mass seemed alive—a field of medusa growth—each stem writhing and curling and twisting of its own volition, in its own particular way, and yet the whole ebbing and flowing as one frond in obedience to the rhythmic breeze. It was the old story over again of the single corpuscle tumbling and rolling individually while yet helpless in the general current of the blood; and of the colonial organism—each individual ant doing his own work although bound irrevocably to the will of the whole, and—who knows—it is perhaps no whit different from the apparent free-will personalities of our separate selves, compared with the destiny of the human race.

I sat me down on a couch of golden, blowing weed, with beautiful green-armed starfish sprawled here and there, and leaning back, watched the bubbles of my life's breath tumble out from beneath my arms and shoulders. From invisibility, from the colorless, formless stream of gas flowing down the length of black hose, they became definite spheres, painted and splashed with all the colors in sight. Once, when I was making my first flight in a plane, I had, for a short space of time, the soul-devastating sensation of being suspended motionless in the ether while the earth dropped away from me. That has never been repeated, but here on the bottom of the sea, looking upward at the great bubbles of breath, I can often conjure up the belief that I am actually looking at a constellation, a galaxy of worlds and stars, rolling majestically through the invisible ether. The background is as mysteriously colorless and formless as space itself must be, and as I peer out through my little rectangular windows I seem to be actually living an experience which only the genius of a Verne or a Wells can imagine into words. It suddenly flashes over me that in giving over my moon and stellar longings for the depths of the sea, I have in a manner achieved both.

I have even the sensations of a god, for in each of the spheres I have created, I see very distinctly my own image. But I also see many more interesting things and my moonings in the present instance were brought to an abrupt end by a glint of gold which appeared on each globule of air—a fiery pin-point which became an oval and soon a great spot as if a sun were rising behind me. If I were looking at a real planet such a thing might be a tremendous volcanic eruption on the surface. Twisting slightly and peering obliquely through

my little periscope I saw what, after all, is the most
joyous thing in life, an old friend in a new guise—
another great golden grouper just behind me, revealed by his reflected image on my ascending
breath.

To my left the rope from the anchor weight led
up in a graceful curve to the distant, dark silhou-
ette of the boat. Now and then a window opened
in the ruffled ceiling and framed the anxious face
of my faithful assistant peering down, on the lookout for approaching danger. The face vanished,
the window slammed shut as the water-glass was
withdrawn, and I was again visually lost to the
upper world.

Two small, black forms approached from the off-
shore side of my aquatic sky, looking from below,
like the keels of funny, diminutive tug-boats, and
driven by a pair of most efficient propellers. These
were rather turbines of sorts, furling and unfurling
in a curling, spiral manner, which offered the most
and the least resistance respectively to the water.
Long rudder tails, two slender, sharp beaks and
sinuous snaky necks came into view, and a swirl
sent both birds into my world—meaning complete
submersion for them. There followed a chase which
no man's eyes have ever seen before—a pair of
flightless cormorants pursuing a scarlet sea bass—
viewed from below. The fish saw them coming and
fled at full speed, not in a straight line but in a
series of zigzags, perhaps, like a chased hen, seeing the pursuers first out of one eye on one side,
then out of the other apparently on that side. The

cormorants separated, one diving deeply while the other followed its prey directly. Soon the confused fish dived at right angles and before it had time to turn again was in the beak of the second bird. The moment it was captured, both birds relaxed every muscle and with dangling wings and feet let themselves be drawn up to the surface. There, even from my depth, I watched a second race begin, and surmised the details of what I had seen enacted twice the day before from the boat, a cormorant coming up with a fish and instantly chased by another, both travelling at such high speed, that with wings spattering and feet going, their entire bodies were almost out of water. At the first opportunity, a quick upward toss, reversing the fish, and a gulp, and down it went headfirst. On this occasion I saw only the frantic disturbance of the surface, rapid dodging, and then cessation of motion, after which the leading bird immersed and shook its beak in the water several times, and I knew that if I so chose, I could write in my journal that at Narborough, *Nannopterum harrisii* includes *Paranthias furcifer* in its articles of diet.

The surface ripples had hardly ceased when a cloud drifted across my little sky. And, parenthetically, at this place I digress long enough to make a certain point clear. As I ramble on of the adventures and sights which came to me in my underworld, there would seem to occur almost a rhythmic succession of happenings, one after the other, like the feats of circus performers who wait in the wings for their turn to come. This works a hopeless injustice to this water world.   Please remember that the exigencies of my place in that world, and the physical makeup of my helmet enables me to see only the merest fraction of occurrences even in an acute-angled single direction. A horse with blinders is a reasonable simile, or better still, an aged, half blind old man, crippled with rheumatism and palsy and dropped suddenly without warning into the busiest of a city's streets and requested to narrate the happenings about him, and give to them some sort of explanation!

Now again, the ripples of the surface above me had scarcely died away to the usual heaving, opaque, moonstone appearance of my water sky, when a cloud came drifting past. If I had been looking behind me some time before, and had eyes which could penetrate the wall of blueness in the distance, this cloud might at first have seemed no bigger than a man's hand. Overhead, however, it was large enough to darken the whole bottom, and, except along the rim, formed a solid mass. At least twenty thousand slender little Galápagos snappers floated over and around me. They were only two to three inches in length, slender and sinuous, greyish black above, silvery below, with seven or more narrow dark stripes running parallel down the head and body. This was the clear-cut vision I had as the host drifted slowly, almost without individual movement, toward and over me. Some danger, forever unknown to me, wrought a whirlwind in this living cloud, and instantly every fish vanished,—the whole becoming a mass of blurred

lines, a great grey something out of focus. As
quickly, fear passed, and every fish again became
clearly etched in its place among its thousands of
fellows. Slowly all passed from view, a few
hundreds along the lower edge sifting through the
uppermost fringe of weeds. It occurred to me
then that their man-given name was a singularly
appropriate one—*Xenocys,*—strange! swift! It
should have been *Xenocys xenocys;* they were too
delicate, too immaterial for any noun.

My sea-lion returned for a last look but slewed
off, and then a turtle, almost as long as myself,
swam into my ken. He was a much more satisfactory constellation than any in the heavens, of most
of which I have never been able to make head or
tail. But he was a turtle at its best. Until one has
looked up and seen eight hundred pounds of sea
turtle floating lightly as a thistledown overhead,
balanced so exactly between bottom and surface
that the slightest half-inch ripple of flipper motion
was sufficient to turn the great mass partly over
and send it ahead a yard—until then one has never
really seen a turtle. Two years ago when I visited
these islands, I watched the little penguins waddling about with their ever inimitable gait, I saw
the cormorants awkwardly climbing over land,
even hauling themselves along by means of crook-
ing their necks, the sea-lions unlovelily caterpillar-
ing along the ground, and great hulks of turtles
ploughing their way as much through as over the
sand of the beaches. It was now my privilege to
see these same creatures in their chosen element,

graceful, glorified reincarnations of their terrestrial activities.   In all of this I had no false illusions concerning my own relative functioning.   While I have never heard any rumor as to my possessing any grace even at my best, yet on these same islands and beaches I can at least correlate my activity, and I can easily run down any of the creatures which I am discussing.   Whereas here at the sea bottom I sprawl awkwardly, clutching at waving weeds to keep from being washed away by the gentle swell, peering out of a metal case infinitely more ugly than the turtle's head and superior to them only in my hearty admiration of their perfect coördination in an exquisitely adapted environment.

My pleasant turtle friend still floated motionless, when suddenly he was the means of my making a delightful discovery in Einstein relativity,—making clear the fact that he was motionless and yet not motionless.   I was resting lightly on a bed of weeds with a generous tuft of them in each hand. I was aware that with every surge there was a very decided movement of the whole mass but as everything in sight was equally shifted my mind registered no definite motion.   Of one thing only was I certain, that however we plants and organisms at the bottom were blowing and vibrating back and forth, the turtle at least, isolated in mid-water, was as still as the distant rocks themselves.   Becoming cramped I decided to stand upright for a while, and gently lowered my feet until I felt them fit into convenient crevices of the concealed rocks

beneath me. This gave me safe anchorage, and in
a minute more all my surroundings, my whole
world, went trailing off as far as it could, then,
with equal unanimity, all faithfully returned. I
glanced upward and was as astonished as if when
on land I should suddenly see the moon or sun begin bobbing back and forth in the sky, for my
turtle was behaving like everything else and being
swayed back and forth, suspended in the invisible
medium exactly as we at the bottom. To look
back upon it, no more silly lack of reasoning could
be imagined on my part, but when you leave the
world for which God made you and wilfully enter
other strange ones, it is reasonable to suppose that
your senses and brain have to become readjusted as
well as your more physical being. For five minutes
I derived infinite delight in alternately swaying
with the weed, and holding to the rock, and thereby
at will giving to my turtle absolute stability or
rhythmical swaying through space. He seemed
quite unaffected by the theory, but fascinated by
the sight of this strange copper-headed, white-
skinned, worm-like being, with an enormously long,
curving tentacle from the tip of its nose, forever
pouring forth a mass of white, bubbly gas, and
which idiotically kept standing up and sitting down.
Never for an instant did the great chelonian take
his eyes from me. If I could put down what he
actually thought of me no halting words of mine
would be necessary in this chapter.

And still the turtle hung in the sky when two
penguins arrived. For a time they swam around in

little intersecting circles, constantly plunging their heads beneath the water to stare at me. Finally curiosity overcame them, they could stand it no longer, and down they came, clad in mantles of silvery bubble sheen. They encircled me once, started on another round but then became fascinated by the black hose and after an examination, half paddled, half drifted to the surface and were gone.

Two mighty schools of *Xesurus* passed me grazing slowly. When within six feet they left off their eternal feeding and formed up into more or less orderly ranks which flowed like some enormously long sea-serpent around the identical corners of rocks where had passed the leaders, yards and yards in advance. Invariably the formation of an irregular line led very close to me, the closing up of ranks evidently being connected with the presence of danger or at least something suspicious or strange. It was an amusing sensation to have these hundreds of fish file past, all rolling their eyes at me as they went. I felt almost embarrassed at times, as perhaps "the remains" must occasionally feel as the viewing crowds stream past. With these yellow-tailed cows were widely scattered, single individuals of a fish which we never caught nor identified. In shape and in the general greyish blue color of body they bore a considerable resemblance to *Xesurus,* their characteristic marks being two white spots above the eyes. But they were not grazers, nor even, I believe, herbivorous. I never saw them graze even when the school of their asso-

ciates remained in one spot, doing nothing else for
a half hour but scrape the algæ from the rocks.
Once too, I saw one of these white-spotted chaps
pursue a small fish, and though he did not capture
it, yet I could not mistake his intent,—there was
nothing of play nor yet of sudden anger in the
attempt, but a very evident desire for food. They
were much more timorous than the yellow-tailed
surgeon fish and at any hint of danger would dart
into the thick of the school. All this makes me
think that they are very likely examples of real
mimicry, gaining a good percentage of immunity
by the resemblance to and close association with
fish, which by their great numbers and poisonous
spines are well able to fight off ordinary dangers.

When I rolled over and looked about, there came
to me a vision of the abundance of life in the sea.
The cloud of little fishes had gone, even the ubiqui-
tous yellow-tailed surgeons were out of sight for
once, and yet from where I sat I could see not fewer
than seven or eight hundred fish, not counting the
wrasse and gobies which played around my fingers
as thickly as grasshoppers in a hay-field. Out of
the blue-green distance or up from frond-draped
depths good-sized grey sharks appeared now and
then. Two came slowly toward me, closer with the
in-surge and then floating farther off with the out-
swing. They turned first one, then the other,
yellow, cat-like eye toward me, and after a good
look veered off. Near to them were playing round-
headed pigfish, a few *Xesurus* swam still nearer,
and even small scarlet snappers, the prey of al-

most every hungry fish or aquatic bird, even these went by without any show of nervousness. The pair of sharks passed on, almost unnoticed, and all the mass of life of this wonder world seemed going smoothly and undisturbed. Far away in the dim distance one of the sharks appeared again, or it may have been another—when, looking around me, I saw every fish vanishing. While I have mentioned what must seem an identical occurrence before, yet this was as different as a great battle is from a street accident. Through copper and glass and air I sensed some peril very unlike the former reaction to the sea-lion, and I rapidly climbed a half dozen rungs, swallowing hard as I went to adjust to the new altitude. Clinging close to the ladder I looked everywhere, but saw nothing but waving seaweed. The distant shark had vanished together with all the hosts of fish, even to the bullying, fearless groupers. I was the only living being except the starfish and the tiny waving heads of the hydroids which grew in clusters among the thinner growths of weed, as violets appear amidst high grass. Whether the distant shark was of some different, very dreaded kind, or whether some still more inimical thing had appeared—fearful even to the strange shark, I shall never know. Five minutes later, fear had again passed, and life, not death, was dominant (Fig. 69).

I climbed to the surface at last, my teeth chattering from the prolonged immersion. This water, although in no sense the Humboldt Current, is much cooler than that at Cocos and I become numb and chilled without knowing it. Excitement and concentrated interest keep me keyed up, and the constant need of balance requires that every muscle is taut, and then when I reach the surface and relax, the chill seems to enter my very bones. Fortunately there is always either rowing or pumping to do and this soon warms me.

During my last dive I had noticed five or six new species of fish and hoping to hook some of the smaller ones I decided to get some bait. I had the boat backed near shore and at a propitious moment on the crest of one of the lesser swells I leaped off. The scarlet crabs here are remarkably tame, far more so than on any of the other islands, a fact for which I can in no way account. The casual visits of man may be of course ruled out as having nothing to do with it, and yet here birds and fish, the crabs' most deadly enemies, are unusually abundant.

With two big, scarlet crabs I vaulted back on the crest of another convenient little swell, fortunately just avoiding the succeeding three, any one of which would have tossed our cockle-shell high up on the jagged lava. I found to my disappointment that we had between us only one hook and that a large one. However, I anchored again near the spot where I had last dived and threw over the hook. I immediately caught one of the round-headed pig-fish, about a foot in length. As I was pulling in a second one, a six-foot shark swung toward him and this gave me a hint upon which I acted at once. I pulled in the fish quickly and studied the situation

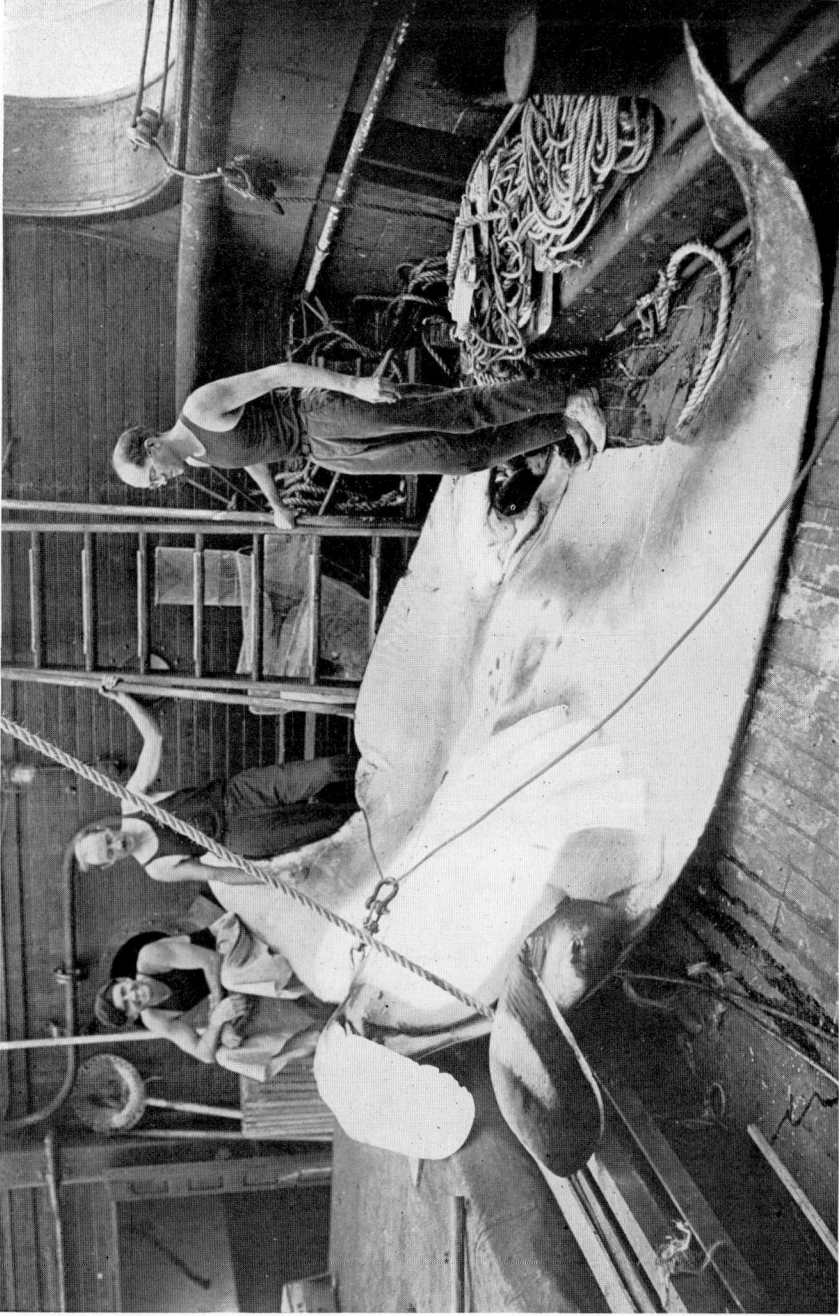

Fig. 30.—Manta or Giant Ray Captured by Dickerman, Franklin and Cady.

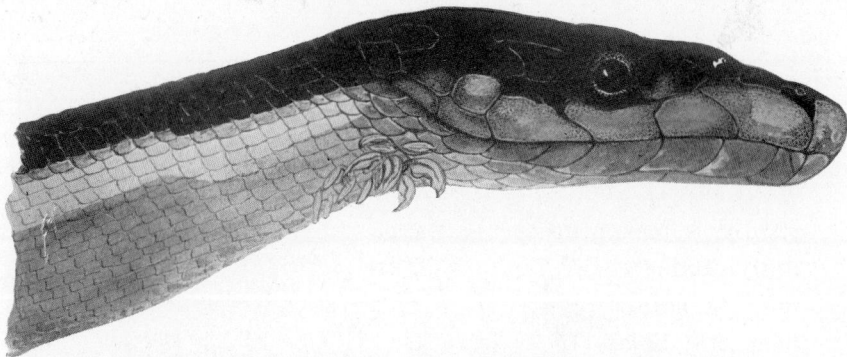

FIG. 31.—HEAD OF SEA SNAKE, FROM THE CURRENT RIP.

Showing a patch of barnacles growing on the skin.

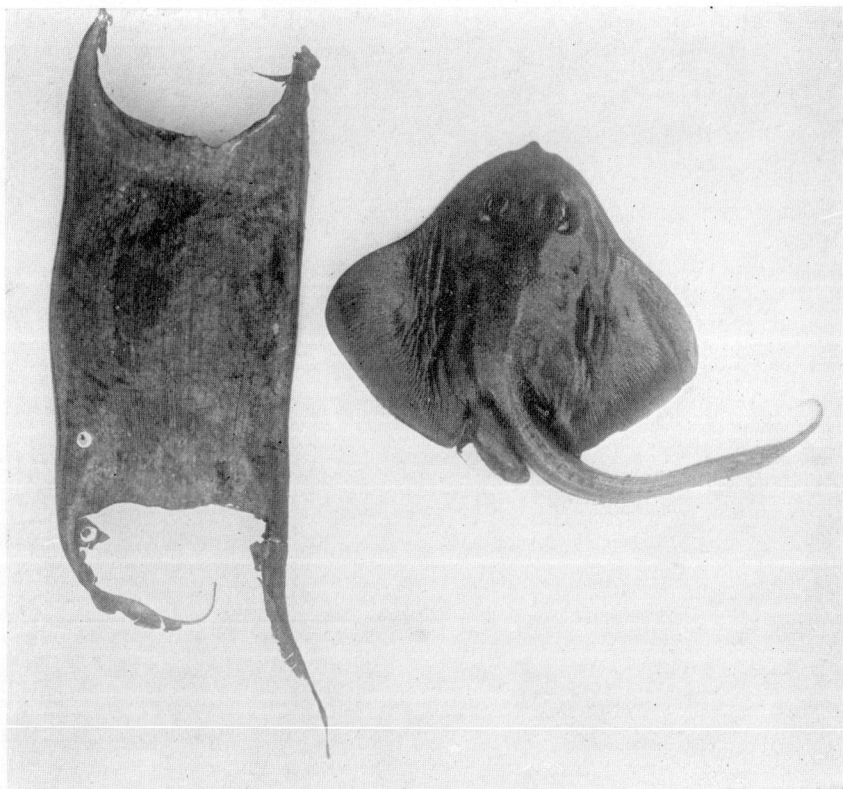

FIG. 32.—EGG-CASE OF A DEEP SEA RAY, AND THE NEARLY HATCHED EMBRYO
IT CONTAINED.

From the bottom nearly a mile down at Station 74.

through the water glass. Two sharks were swimming slowly about the very rock where I had been sitting a few minutes before, probably the same individuals who had then been so curious about me. A small group of the pigfish swam around, over and below the sharks, as they had also done when I was submerged, sometimes passing within a foot of the sharks' mouths without the slightest show of emotion, of fear or otherwise. An angelfish and two yellow-tailed cows passed, and a golden grouper together with two deep green giants of the same species, milled around beneath the boat, cocking their eyes up at us, now and then.

I baited the hook with a toothsome bit of crab and lowered it. All the pigfish rushed it at once, and as it descended, the sharks and groupers followed with mild interest, almost brushing against it, but wary of the line. Failing to elicit any more practical attention from the golden grouper I allowed one of the pigfish to take the bait and hook. Then, watching very carefully, I checked his downward rush, and swung him upward. He struggled fiercely and like an electric shock every shark and grouper turned toward him. Without being able to itemize any definite series of altered swimming actions, something radical had happened. The remainder of the school of pigfish, while they remained in the neighborhood, yet gathered together in a group and milled slowly in a small circle. There was no question that from being a quiet, slowly swimming, casually interested lot of fish, the three groups—pigfish, groupers and sharks—had become sur-

charged with interest focussed on the fish in trouble.
I drew the hooked fish close to the boat, and could
plainly see that the hook had passed only around
the horny maxillary. There was not a drop of
blood in the water, and the disability of the fish consisted only in its attachment to the line. Yet the
very instant the struggle to free itself began, the
groupers and sharks, from being at least in appearance friendly, or certainly wholly disregarding the
pigfish, became concertedly inimical, focussed upon
it with the most hostile feeling of an enemy and its
prey.

For half an hour I played upon this reaction and
learned more than I had ever seen or read of the
attacking and feeding habits of groupers and
sharks. When the struggling began the sharks all
turned toward the hooked fish. Not only the one
nearest who must easily have seen it for himself,
but two, far off, turned at the same instant, and
within a few seconds two more from quite invisible
distances and different directions. What I saw
seemed to prove conclusively that sharks, like vultures, watch one another and know at once when
prey has been sighted by one of their fellows. The
numerous sharks thus call one another all unin-
tentionally, as when one of our party caught a
shark at Cocos, and in an incredibly short time there
were seventeen attacking it. On the other hand it
must be admitted that sharks differ from vultures as widely as the poles in the matter of scent.
Vultures all but lack this sense, while we know that
fish have it well developed. But even in the case

of blood in the water, it seems to me that diffusion
cannot be nearly rapid enough to account for the
instantaneous reaction on sharks near and far. The
phenomenon is as remarkable in general aspect as
the apparent materialization from the air of a host
of vultures where a few minutes before none were
visible.

Even more than in this problem, I was interested in the exact method of feeding of sharks and
groupers. After making sure of the first phase of
interest, I allowed a six-foot shark to approach the
hooked pigfish. It came rather slowly, then with
increased speed and finally made an ineffectual
snap at the fish. The third time it seized it by the
tail and with a strong sideways twist of the whole
body, tore the piece off. The second fish attacked
was pulled off the hook, and two sharks then made
a simultaneous rush at it. So awkward were they
that one caught his jaw in the other's teeth and for
a moment both swished about in a vortex of foam
at the side of the boat.

I noted carefully about thirty distinct efforts
or attacks on the hooked fish, and only three times
was I able by manœuvering the fish to get the
shark to turn even sideways, never once on its back
as the books so glibly relate. I sacrificed seven pigfish, and then tried to get the golden grouper but
it was too wary. A giant five-foot green grouper,
larger than any we had taken thus far, was becoming more and more excited however, and when I had
tolled him close to the surface I let my fish lure
drift loosely. One swift snap and the entire fish

disappeared, then a single slight nod of the head and the line parted cleanly. The general effect was of much greater force and power exerted in a short space of time than in the case of the sharks. When it comes to lasting power for only a short time, after being landed, however, the groupers fight while the sharks smash and thrash until they are actually cut to pieces.

After this exhibition, without hesitation, I dived in the helmet again in this very spot with no change in the attitude of the sharks toward me. I had had these sharks close to me a little while before, and although my efforts under water seem to me no less awkward and helpless than a hooked pigfish, yet to these so-called man-eaters, there is apparently all the difference in the world, and I was absolutely safe from attack.

Mr. Zane Grey, who, at my recommendation, went to Cocos and the Galápagos, had as his object big-game fishing, and as the following paragraphs will show, he underwent the same experience that we had, both when we were here two years ago on the *Noma,* and now again on the *Arcturus.*

Fishing off Chatham Bay, Cocos Island, he writes in his book "Tales of Fishing Virgin Seas," —"The next hour was so full of fish that I could never tell actually what did happen. We had hold of some big crevalle, and at least one enormous yellow-tail, perhaps seventy-five pounds. But the instant we hooked one, great swift gray and yellow-green shadows appeared out of obscurity. We never got a fish near the boat. Such angling got on my nerves. It was a marvellous sight to peer down into that exquisitely clear water and see fish as thickly laid as fence pickets, and the deeper down the larger they showed. All kinds of fish lived together down there. We saw yellow-tail and amber-jack swim among the sharks as if they were all friendly. But the instant we hooked a poor luckless fish he was set upon by these voracious monsters and devoured. They fought like wolves. Whenever the blood of a fish discolored the water these sharks seemed to grow frantic. They appeared on all sides, as if by magic.

"By and by we had sharks of all sizes swimming round under our boat. One appeared to be about twelve feet long or more, and big as a barrel. There were only two kinds, the yellow sharp-nosed species, and the bronze shark with black fins, silver-edged. He was almost as grand as a swordfish.

"While trying to get the big fellow to take a bait I hooked and whipped three of this bunch, the largest one being about two hundred and fifty pounds. It did not take me long to whip them, once I got a hook into their hideous jaws. The largest, however, did not get to my bait.

"An interesting and grewsome sight was presented when Bob, after dismembering one I had caught, tumbled the bloody carcass back into the water. It sank. A cloud of blood spread like smoke. Then I watched a performance that beggared description. Sharks came thick upon the scene from everywhere. Some far down seemed as long as our boat. They massed around the carcass of their slain comrade, and a terrible battle ensued. Such swift action, such ferocity, such unparalleled instinct to kill and eat! But this was a tropic sea, with water at eighty-five degrees, where life is so intensely developed. Slowly that yellow, flashing, churning mass of sharks faded into the green depths."

Again Zane Grey writes from Darwin Bay, Tower Island, after hooking a huge shark: "Then the fun began. It really was not fun, but work under a hot sun, in a bobbing boat, with thundering surf always threateningly near at hand, and most unforgettable of all, with a school of huge black sharks following the one I had on. When I got the double line over the reel I kept it there, and as a consequence had the shark in sight all the time. His comrades glided between him and me, bumped the boat with their tails, and acted in every way to convince a reasonable angler of their dangerous mood. They were undoubtedly man-eating sharks. If R. C. had not been in sight and within call I never would have risked my life in that cockleshell of a launch, amidst a swarm of ravenous wolves of the sea. At length this one, like the other two, broke my leader, demonstrating fully that this especial kind of copper wire was useless for fishing."

Now Mr. Grey is probably the foremost big-game fisherman of the world, and knows more of the habits of these fish from the sportsman's angle than any of his fellow human beings. Under the circumstances that he describes, few men, certainly

PARTI-COLORED BUMPHEADS
Bodianus eclancheri (VALENCIENNES)
(Average length fifteen inches.)

PARTI-COLORED BUMPHEADS
Bodianus eclancheri (VALENCIENNES)

(Average length fifteen inches)

PLATE IV

not I for one, would have dared to think otherwise than he did of the sharks of Darwin Bay. And yet, after all, their man-eating, dangerous qualities were circumstantial, and engendered by what he observed in their attacks on hooked fish, of their own or other species.

Less than a month after he left this wonderful bay, the *Arcturus* anchored in it, and a few days thereafter Dr. Gregory, Ruth Rose, myself and all the rest of my staff were diving in helmets, and walking about the bottom, with these self-same "man-eating" sharks swimming by and around and over us, dashing at and taking our hooked fish, but, except for a mild curiosity, paying no attention to ourselves. It was as unexpected to me as to any-one, yet I will go on record as saying that it is per-fectly safe to sit or walk around, or climb up and down ladders and ropes, to leap or twist quickly about, or to sit motionless, protected only by a copper helmet and a bathing suit, among the sharks of Cocos and the Galápagos, whether they are swimming slowly along, or devouring some fish, dead or in obvious trouble.

CHAPTER VIII

FLOTSAM AND JETSAM

If heat is the mother of all life then water is surely its father. We came from the water, we are still absolutely dependent upon it, two-thirds of our entire body is nothing but water. In our physical frame we carry with us many aquatic memories, water-logged characters which point to distant amphibious or submarine ancestors. The mark of the sea is upon us though our home may be in the heart of a continent.

The simplest of beings are inhabitants of water —mere droplets of movement, hesitant on the threshold of life, as yet neither quite plants nor animals. In comparison, a forming crystal may seem a great advance, a restless oil globule suggests a sentient organism. But the droplet of life can afford to rest motionless. It treasures in its minute nucleus a something possessed by neither crystal nor globule.

It would almost seem as if water, especially sea water, had some slumbering force within itself, a dormant sympathy for organic life which needed merely the slightest stimulus to awaken and to take its share in dynamic animation. A suspended cobweb vivifying the air about it into complex activities would be no more of a marvel than the jellyfish which moves through the sea and is itself the very essence of water. Dry it, and there remain neither bone nor tendons, disturbed organs nor traces of blood, but only the faintest of glistening films, which disintegrates and blows away with the first breath of air. Yet imbued with its ninety and nine parts of salt water, it moves and contracts and throws its poisoned darts, it swallows and digests, and dimly sees and feels, it produces eggs and strews them like chaff as it slowly vibrates on its course. Yet so evanescent is it that it seems like some organic mirage. The eye often misses it altogether, looking straight on and through its being, and finally locating it by its shadow. The earth-wide basins of liquid gently sustain and capably support the host of beings who experience life and death among the waves. In countless ways each tiny creature is ministered to, and given his chance to fight upward toward the unknown caste-to-come which seems the sole object of the existing of these lives.

Important as water is to all higher creatures, its actual astounding percentage in tissues and organs is more and more completely concealed from view. But always we perceive new, unexpected qualities. And when unusual demands are made they too are granted. Creeping upon the mud and coral are myriads of shellfish whose flesh would tempt every passing fish. So when their need cries aloud for protection, the Father of Life comes to their aid. By some strange, secret alchemy they draw from the transparent water the hardest and most durable of walls, and encase themselves in shells of lime, of marvellous architecture and splendor of pattern and pigment.

In the course of past time, fishes of the sea covered themselves with scales of shining silver and developed four important fins—prophecies of wonderful legs and arms and feet and hands, if one could only have known. But in those times the Great Father of Waters was in no fear about the desertion of his children. Fishes leaped from the waves and even learned to skim through the air on outstretched fins. But they always plopped back exhausted. And when other creatures insisted on clambering out on mud-banks and flipping themselves along, the great breakers merely chased them and good-naturedly rolled and tumbled them back again into the green frothy water. And the ocean in those days swept round and partly over the half dried land, and the sound of the storm waves vibrated uselessly around the headlands and through the valleys, for there were no ears to hear.

By the time the first little monkey climbed down a swaying vine for his evening's drink, the dominion of the sea had become lost in the past. The earth was galloped over and burrowed into by myriads of beings; trees were perched on and bored through; the air hummed and whistled with wings and webs and leaping forms. So completely a thing of the past had the sea life become that many creatures had gone back to it as to quite a

new element. Their old, old aquatic memories
helped them not at all, and the penguins had to
re-stiffen their feathers into scales, and to encase
their wings in immobile mittens cut after the fashion of sharks' fins. And the seals ceased the running about upon the land and became completely
readapted to a sea life.

So let us return, at least mentally, to the Sea,
for there is no happening on land which cannot
there be duplicated and often bettered. But to
appreciate these similarities to the full, one must
become amphibious. As well live in Kansas or
Switzerland and know the ocean only in the ency-
clopedia volume MUN to ODE, as sit in a deck
chair and watch it pass or scan its waves with
binoculars. To such a watcher no real secret is
ever confided—he thinks in terms of waves and
swells, and his eye is held by the horizon beyond
which is the dry earth for which he longs. But
to the aquatic devotee, the oceanic fan, surprise
after surprise is vouchsafed, for to him the three
elements are not phenomena wholly apart.

We are grateful to the dry land for standing
room, to the air for the breath of life. But any
glance askance at the watery depths is but a pitiful or a comic gesture when we remember that
85% of our brain is water, and much more akin
to salt than to fresh. To be sure we cannot drink
salt water and live, but when necessary it is an
admirable temporary substitute for blood itself,
whereas sweet water would be a fair poison in our
veins. Take the man who shudders at the thought

of the ocean's depths, and put him in the midst of
a tropical desert at breathless noon, or make him
climb the Himalayan hills until his very marrow
is frosted with the winds which caress Kinchin-
junga, and his lungs cry out for their need of
oxygen,—and his natal earth will seem quite as
inimical as the great waves of mid-ocean or the
black liquid depths.

For countless voyages I have hung over the
bow of passenger steamers in mid-ocean, making
of myself a figurehead of sorts, straining my eyes
downward to watch the living creatures which
whirled into sight and swept past. Dolphins, flyingfish, tunny, an occasional shark—these are
familiar to all who have ever glanced over the
bow. But the rays of the slanting sun striking
obliquely into the smooth surface often revealed a
myriad, myriad motes—more like aquatic dust than
individual organisms, which filled the water from the
very surface to as deep as the eye could penetrate.

Toward sunset these would vanish in the increasing dimness, and finally the bow would cut
its way through an opaque, oxidized liquid, as
unlike water as tar to glass. The moon overhead
which showed in the waning day as a crescent of
cloud, now cuts through the darkness like a sliver
of gold. So the minute sea life becomes, in the
dark, redoubly visible, and the ship ploughs a deep
furrow through miles of star dust—phosphores-
cence which will fill the last imaginative human
being as full of wonder and awe as it did the first
who ever ventured out to sea.

As I have elsewhere explained, the floating oceanic life is known as plankton—indicating the helplessness of these wanderers, drifting about at the direction of the winds and currents. Even vaguely to estimate the abundance or numbers of these powdery clouds of animals of the ocean is to attempt a Herculean task, second only to numbering the sands of the shore or the proverbial hairs of our head. One dark, moonless evening I put out a silk surface net the mouth of which was round, and about a metre or a yard in diameter. At the farther end of the net a quart preserve jar was tied to receive and hold any small creatures which might be caught as the net was drawn slowly along the surface of the water. This was done at the speed of two knots and kept up for the duration of one hour. When drawn in, the net sagged heavily and we poured out an overflowing mass of rich pink jelly into a white flat tray. This I weighed carefully and then took, as exactly as possible, a one-hundred-and-fiftieth portion. I began to go over this but soon became discouraged, and again divided it and set to work on one sixth of the fraction on which I had first started. After many hours of eye-straining and counting under the microscope, I conservatively estimated my 1/150 part of the hour's plankton haul as follows:

Feathery copepods—Candace-like	7,920
Bright blue copepods—Pontella-like	71,400
Other copepods—Calanus-like, pink	139,320

Bivalve crustacea—Ostracod-like	4,920
Short-eyed shrimps	720
Siphonophores	14,400
Helix snails	8,880
Purple Ianthina snails	13,440
Egg masses of snails	1,080
Free eggs, various	5,280
Arrow-like flying snails	2,520
Nautilus-like flying snails	240
Oyster-like flying snails	960
	271,080

If we multiply this by one hundred and fifty we get forty million, six hundred and sixty-two thousand individuals. Please remember that this is a very conservative estimate of only a few of the more easily counted groups in one small haul of an hour's duration, and the magnitude of the life of the sea will begin to dawn upon our minds. Twelve hours later—in full daylight—I repeated the haul as closely as possible and, instead of forty million, I captured about one thousand individuals of the corresponding groups. So although plankton is an involuntary horizontal wanderer, yet vertically it has more perfect control, and having developed its own system of lighting it will have nothing of the sun or even of moonlight, and remains well below reach of the stronger rays.

My own interest in plankton is wholly that of trying to disentangle the lives of some of the small people—to put myself in their places by day and night, but I feel that I must establish their im-

portance in the minds of more practical and far-
seeing readers.   Realize then, that even for our
human race, the universe of plankton is of vital
importance.   The surface-loving copepods are
commonly and correctly known as "whale food,"
and they are also the most important food of many
fishes.   Only at the surface can vegetable life exist
and develop, changing sunlight into edible mater-
ials, and in plankton diatoms and other plants
affording satisfactory aquatic fodder to the small
grazing animals about them.   They thus start the
ball of life rolling, which does not cease until it
includes the possibility of continued existence for
whales and food fishes, while, in the future, the
whole human race may come to depend upon this
larder of ocean.

Indeed it is a remarkable fact that ship-wrecked
men in an open boat, if their lot is cast on waters
rich in plankton, need never starve to death if they
can manage to drag an old shirt, net fashion,
through the water at night.   The great percentage
of crustaceans makes plankton a rich, nourishing
food, even raw.

I can imagine no swifter way of killing anyone's
interest in plankton than to put him in front of
a pan of forty million swarming small folk.   We
have only a sort of hypnotic or at most superficial interest in a regiment or mob; and so I
gave but the merest mechanical attention to the
thirteen thousand odd *Ianthina* snails in my counting tray.   But when I lay flat on my pulpit platform and began scooping up, one by one, the

creatures which for years I had watched go past out of reach, then my distant longings began to change into intimate acquaintanceships, and I learned to admire and to have a real affection for these little fellow beings who lived their lives with me on this whirling planet.

Hummingbirds vibrate before flowers, albatrosses skim for hour after hour over the waves, but sooner or later every bird must come to rest—its muscles exhausted, its wings aweary. But for the mid-ocean folk there is no rest as we know it. Somehow or other they must keep themselves suspended. A list of possible ways, thinking as always from our own experience, would include swimming or flying, treading water, balloons of air or gas, or clinging to some bit of floating wreckage, whether from a storm-broken ship, a bit of porous lava, or a pinion dropped by a passing seabird. All these and many others are actually in use, and had been so for millions of years before man had brain enough to make a list.

Oblong pieces of whitish scum had tantalized me for many voyages, and even when I had emptied one of these bits from my net into a small aquarium I could make nothing of the mass of bubbles, until I looked beneath the surface and there saw that exquisite violet sea-shell with the euphonious name of *Ianthina*. Although this snail lives in a home of tissue-thin lime, it yet spends its entire life at the surface of the ocean. Its relations which we know on land leave a trail of glairy slime wherever they walk, and *Ianthina*

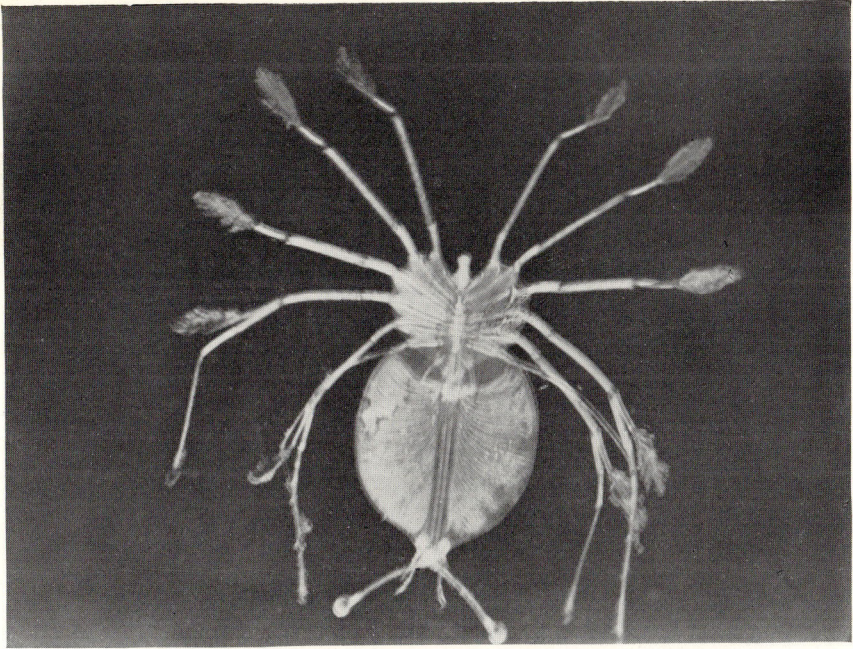

Fig. 33.—Phyllosoma—A Transparent, Larval Crustacean.

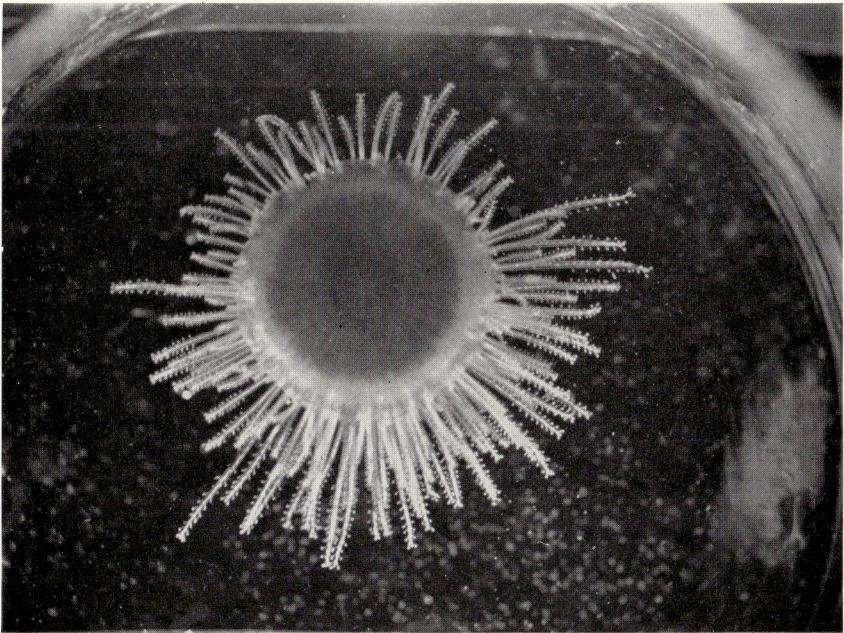

Fig. 34.—*Porpita*, One of the Most Beautiful of the Floating Jelly-Fish.

FIG. 35.—GIANT RAY OR DEVILFISH SWIMMING AT THE SURFACE.

This was harpooned later, and measured eighteen feet from tip to tip of the wing-like f

still has the gland which secretes this, but has
etherealized its use. The thin secretion is poured
forth, and then, by successive upreachings of a
part of the foot, bubbles of air are caught and entangled in the slime, which soon extends out as a
narrow buoyant raft, the shell hanging down at
one end. The bubble slime is not only balloon but
nursery, and egg after egg is suspended from the
lower surface. So abundant were these snails that
I observed them with only general interest, thinking of course that their whole life history was
well known, but on my return I found that this was
far from the case, and that few facts are known
about them.

There are two kinds of thrills in science; one
is the result of long, patient, intellectual study.
An example of this is the years of astronomical
calculation whereby movements of certain heavenly
bodies can be explained only by the existence of
some unknown factor, and then one day this unknown but expected star is found at the very spot
indicated by mathematical necessity.

Another thrill lies in an absolutely unexpected
discovery. Night after night small white spots
floated about on the water just beyond the glare
of the gangway electric lights. In vain we tried
to net them. Now and then several would join
together in a sinuous row and swim slowly along.
At last, with an effort which almost precipitated
him into the sea, Serge Chetyrkin scooped one up
and dropped it into a small jar. To my astonishment I saw it was an argonaut or nautilus—a

paper nautilus—which, in other words, is a diminutive octopus with the most exquisite shell in the world. Never have I seen a creature with a more explosive temper—we named her Mrs. Bang on the spot. Hardly had I changed her to a small aquarium when she angrily shot forth a cloud of sepia, and had to be transferred twice before her ink-bag was emptied and I could observe her clearly.

She rested quietly on the bottom with her many arms wrapped about her beautiful brown and white shell. But as soon as my face approached the glass, she rushed back and forth, shooting directly at me or bumping against the opposite glass, and finally backing into a corner. Here she spitefully squirted spouts of water through her siphon, until I gave her a small fish. She snatched it ungraciously, bit its head off and ate the body, feeling suspiciously about with three or four arms in my direction the while (Fig. 15).

Two days later she went into such a paroxysm of rage that she flung herself clear out of her shell. I carefully picked this up and found her eggs still remaining inside. There were thirteen hundred of them, even-ended ovals, about ten by fifteen millimeters, with a tiny thread at one end which attached them loosely together, exactly like a miniature bunch of grapes—the smaller stems growing out from larger and these in turn from a twisted, central rope. The embryos were in various, well-advanced stages, with the future eyes of the infant argonauts marked by two large, red spots.

The shell of the argonaut is secreted by two

great flat plates on the arms, and it was formerly thought that when, in calm weather, the owner rose to the surface, it sat back comfortably in its shell, raised the two broad arms aloft and used them as sails. Such a performance should properly take place only within sight of the fleets of entangled ships in the Sargasso Sea!

I never tired of watching the squids and octopuses which we captured. Soon after we landed the nautilus, Serge, with his usual skill, caught a two-foot squid which I studied for many minutes. It squirted sepia all over us and bit our hands before we could drop it in an aquarium. When it quieted down it pulsated slowly, while the colors came and went over the body in such a way that new adjectives will have to be coined adequately to describe it,—reds, blacks, browns, yellows, rolling, surging, springing into vision as the pigment spots contracted or expanded, a living, liquid palette.

The staring eyes were oval, and of an astonishing turquoise blue, and even on this surface, scarlet spots grew and passed—vanishing completely, only to reappear and coalesce so that the turquoise became carnelian. I looked into the sinister, narrow, cat-like pupils, and they seemed to express all the horrible mystery of things which should not be,—such as these monstrous, flabby creatures calling the snail, the slug, the nautilus and the oyster brothers—possessing not even the prestige of having fallen, like the humble sea-squirts, from higher aspirations—shellfish they are and nothing else. And yet unreasonably possessing an eye, as well as or better developed than our own. When to a low evolved mollusk thing, there has been given a "window of the soul" such as this, one wonders what secret, what thing of enormous value must have been bartered for it, what sinister transaction at some nefarious "Bureau d'Echange de Maux." A hand even, would not have been so unexpected, nor a foot patterned after those of infinitely higher beings, but such an eye should not be in such a body.

Before we lose ourselves among the small folk of mid-ocean let us strike a contrast. Day after day, from the crow's nest or the bridge we caught sight of the monsters of the ocean's surface,— occasional sunfish so gigantic that, so long as they remained out of reach of a yard-stick, it were better for a scientist to call them merely exceedingly large. A layman might use the simile of a vertical barn-door and not exceed the truth. Indeed the same time-worn phrase if considered horizontally would be less than the actual fact if applied to some of the devilfish or giant rays which we saw. Now and then a playful one would leap almost out of the water, or pass close to the bow on its graceful, leisurely aquatic flight (Fig. 35).

North of Narborough they were so numerous that three of the staff, Dickerman, Franklin and Cady, made up their minds to capture one. Assembling every weapon, legitimate and otherwise, which the *Arcturus* afforded, they set out in a tiny rowboat and made good. When, later on, we analyzed the fight from the motion pictures, we realized that luck had surely been with us, for if the great fish had slapped its wing tips a little nearer and higher, the rowboat and devilfishers would have been flattened. When once a harpoon was deeply fastened to the fish, the battle became merely a question of trying to tire it out, and to hope that the injury inflicted by the hail of bullets would antedate the effect of their accumulating weight!

Something at last was effective and after two hours the devilfish surrendered and was towed to the *Arcturus*. Several lashings were broken before it was at last drawn out of the water and lowered on the deck (Fig. 30). Here was a specimen indeed, not to be placed on the stage of the microscope, but studied by walking around, over and almost into, for its gaping mouth was quite four feet wide. From fin tip to fin tip it measured exactly eighteen feet, and little by little as we cut it up we weighed the pieces and found it to total two thousand, three hundred and ten pounds. The liver alone weighed as much as a man, and we found a young devilfish about to be born,—a lusty infant weighing twenty-eight pounds and with a fin spread of over three and a half feet. As usual the fish had many interesting parasites. I took eight sucking fish from its gills and at least thirty more fell off when it left the water. On the skin were many weird-looking parasitic crustaceans.

These great fish are not especially wary and a few days before when returning from a diving excursion near shore we played with one for an hour, bumping into it continually with our bow and being splashed by the threshing fin-tips as it half turned over. There were two close together, each with a ten foot expanse of wings. They refused to leave or to go down although we pummeled them with the oars, and they were still swimming and rolling about when we left.

Merely to enumerate the species of floating, living beings which we took in our surface nets would fill this chapter, so all we can do is to think for a moment of the most characteristic ones. If a cupful of pond water is examined, tiny creatures will be seen shooting about, and under the lens one of these resolves into a crustacean thing, with two enormously long horns or antennæ, and a single, median eye. This is aptly named *Cyclops,* and is a member of the group of copepods. We may recall that these little beasts comprised thirty million of our enumerated plankton haul, and so abundant are they that they usually give the characteristic color to the hauls or even to the ocean for miles around, varying from carnelian red to deep madder blue.

Oceanic crustaceans in general and copepods in particular correspond in numbers and variety to the insects among terrestrial creatures. Indeed as regards beauty and variety I can compare copepods only with snow crystals. Very small species often contained good-sized oil globules which seemed to serve the purpose of buoyancy, but these were lacking in larger, bizarre forms who relied on the most amazing development of appendages, some having widespread, feathery tails affording a great expanse of surface for support in this thin medium.

In the dark a small dish of this plankton would glow like a trayful of diamonds, but in the light no trace of luminescence could be detected. And yet, now and then, even under the binoculars there would come a flash as of fire opal. Little by little I narrowed this down until I had in the field of vision a single oval copepod, about an eighth of an inch long. When viewed from the side it showed as a mere tissuey line, but when it turned on its back every color of the spectrum was kindled. *Sapphirina* is its name, but Opalina would be more appropriate.

Traces of Aquarius or Pisces might reasonably be expected in these submarine regions, but hardly of Sagittarius, and yet hardly any pipette of plankton would fail to show numerous little arrows shooting across the field of vision. These are worms in structure if not in conventional outline, but their name *Sagitta* makes up in aptness what they lack in vermiformity. They are transparent, slender and quite stiff, with well-marked fins. The entire anterior end is composed of a mouth armed with great teeth-like bristles, indicating a type of life and diet far different from that of the quiet, plant-eating copepods.

Many of the day-time animals which called the surface of the ocean home, were ultramarine above and silvery white beneath, stained thus with the

very essence of their surroundings,—a vital factor
in helping to hide them from the eye of enemies
which looked down upon them from the air, or
upward from the depths.   But at night a host of
small creatures found safety in being divested of
all pigment.   In the course of evolution they had
scraped off all the mercury from the back of their
beings, becoming so transparent that the food
which they swallowed was the most conspicuous
and opaque part of their anatomy.

I could never quite escape from a decided Alice
in Wonderland feeling when I looked into a dish
of night plankton scooped from the surface.   By
keenest scrutiny I could perceive only the usual
hosts of small fry, when, reaching down and lifting out what seemed only an area of clear water,
there would materialize before my eyes a *Phyllosoma* (Fig. 33).   This was a creature who cast no
more shadow than the thinnest skim of clear ice.
Yet it was a living animal, more than three inches
long, with all the general organs which we ourselves possess,—eyes, mouth, feet, stomach, nerves,
muscles and a strong will to live.   *Phyllosoma,* or
leaf person, was the only name I could give them,
although glass crab would be more appropriate,
for they were the young of some lobster-like
crustacean and nothing is known of the inter-
mediate stages.

On land the barriers which confront animals are
very apparent and tangible—mountains, deep
valleys, rivers, lakes, the presence or absence of
treeless plains, etc.   At sea, living creatures are

confined with almost equal rigidity by invisible walls. Temperature, salinity, pressure and light are some of the intangible and impassable frontiers. But the study of these requires a maximum of diagrams and schedules which would be out of place in this volume. Nevertheless, there is drama and tragedy, plot and adventure, so let us consider sunlight and darkness, or even light and shadow. I have already told how the beings who love the surface of the sea at night are all but absent from it in the daylight, but many others are willing to come up if they can find the merest excuse or parody of a sheltering shadow.

I will work up to concrete examples by a few minutes' observation from the pulpit, which always revealed the life and death need for even the slightest protection. The most faithful attendants of the *Arcturus* were the tunny fish, who kept close to the bow hour after hour, yielding to the occasional dolphins but returning at once when they had gone. Looking down through the ultramarine film I saw a score of these fish metamorphosed to rainbow colors—rich violet bodies with yellow finlets and black tails. Now and then an unfortunate flyingfish rose, then a tunny turned aside, there was a flash in the air of molten silver and the tunny was back. A few minutes later a dense mob of several thousand half-beaks rose like hail. These fish are on their way to becoming flyingfish, and, sculling frantically with tail fins, skim through the air, like planes near the end of their taxiing run. Every tunny within sight flung itself headlong into the boiling mass, took toll, and returned to the pace-making bow race.

Ten minutes more passed and a *Pyrosoma* drifted by—a great, pink, hollow, cylindrical colony of unfortunates who had just missed being vertebrates like the tunny and ourselves. Beneath this cylinder of jelly was a half-dozen pilot fish. For some reason—and this is the crux of the whole matter—so long as they crowded beneath it, no tunny paid any attention to them, although so far as actual concealment went, they might just as well have been hiding beneath mosquito netting or a Greek peristyle. As our bow approached their living roof they became panic stricken. All six little fish dashed out, and as if moved by the same mechanism, six tunnies gave six snaps in the very foam of the bow wave, and six little pilot fish were relieved from further worry about their destiny. It cannot be that the tunny fish do not see their ambushed prey, but as a cat will often wait until a mouse makes some movement before it springs, so there may be some instinctive, hair-trigger, piscine law, of vital moment to them, but which in our own case we would similize with the sporting chance of a wing shot.

I came to have the feeling that far down beyond where my eyes could penetrate were uncounted hosts of little eyes peering upward, waiting for the revealing sunlight to lessen, as animals and flowers appear along the edge of retreating snow, following it, occupying every bare piece of ground. The cook would throw over an empty tin can, and if it failed to sink there would soon be a small fish swimming close beneath it. I could imagine the widening cone of shadow which the can cast downward and the fish, feeling its comfortable darkness, followed it up until it focussed on the bobbing bit of floating tin.

In calm sunny weather as the *Arcturus* steamed along at full speed, few or no fish were to be seen in the open water. Then "full stop" would clang when I decided to sound or take temperatures and soon after we began to float quietly, on the shady side, fish, small crabs and other creatures would begin to collect, coming up from deeper levels into this premature twilight. These, however, were only the skirmishers on the edge of the great nocturnal host—that vast army who could never be fooled by an artificial night and who kept far down below the twilight zone, waiting for the blotting out of the sun before they began their upward rush. I had read of this interesting vertical migration before I started on the *Arcturus* and the contents of every net proved its magnitude. But not until I inaugurated a series of twenty-four-hour surface hauls, taken at fifteen minute or half hour intervals, did I appreciate the clock-like regularity of the movement. After a little practice, I knew that if I wanted a certain type of nocturnal surface fish, a haul at 4.15 to 4.30 A. M. would invariably capture some, while the net drawn from 4.45 to 5 o'clock would never contain a single one.

At Station Seventy-four, I made twenty-four hauls in as many successive hours and took over

four hundred fish of twenty-six species. Up to
6.30 in the evening all the more abundant surface
fish of the daytime were captured, such as pilot
fish, half-beaks, flyingfish and young Seriolas.
After this, not one was ever seen, but promptly at
7 o'clock six species of lantern fish, *Myctophid,*
appeared, and a half hour later their enemies, such
as *Astronesthes,* were taken. In early morning
the reverse occurred, and only one species of lantern fish ever lingered after 4.30 A. M. up to which
time they were taken in dozens (Plate V).

In the case of most oceanic organisms we cannot
tell by a casual examination whether they are
diurnal or nocturnal, but even if we had never seen
a living lantern fish, their equipment of lights,
like that of fireflies, could mean nothing but a life
spent in darkness. This luminescence in sea creatures has always held a great fascination for me,
and when first I saw among a mass of plankton
several of these fish, it was a memorable event,—
like my first electric eel, or my last glimpse of the
Himalayas. My interest in the subject was
whetted when I had translated a recent résumé of
the subject and found that nothing but casual and
fragmentary observations on living luminescent
fish had been made, and these mostly by fishermen.

Several times I rushed to the photographic dark-
room with a dead or dying specimen, to see nothing
but the gleam from the numerals of my wrist
watch. Then one evening I filled a small aquarium
with cool sea water and placed in it three newly
caught Myctophids. Suddenly one of them flashed

LUMINOUS SURFACE FISH IN DAYLIGHT AND DARKNESS

An Eater-of-stars (*Astronesthes*) in pursuit of two
Lantern-fish (*Myctophids*)

Fig. A.   The fish in daylight
" B.   The fish in darkness showing various types of
luminescence.

PLATE V

FIG. A

FIG. B

out so brilliantly that the glass dish, our hands and our faces were clearly outlined. Lin Segal and I spent many evenings in this research and recorded a great number of separate interesting facts, which, like all pioneer work, must be presented in their place without connection or correlation. Out of the mass, however, there are certain ones which fall into orderly relationship, and give a faint but tremendously suggestive hint of the life which these fish lead in the darkness of their underwater world.

I shall consider only the slender-tailed lantern fish (*Myctophum coccoi*) which I took in numbers both in the Atlantic and the Pacific. Imagine a minnow (Colored Plate V), which is iridescent copper above and silvery white below, not over two inches in length, with large eyes and moderate fins. A full-grown fish weighed a gram, which means that it would take about four hundred and fifty to make a pound. It feeds on copepods, sagittæ and other minute plankton fry, and from this food it generates energy to live, to fight, to migrate up and down, to keep illumined one hundred lights and to lay upwards of seventeen hundred eggs.

Scattered over the body are many small, round, luminous organs, which we may divide into three general sets. First, thirty-two ventral lights on each side of the body, extending from the tip of the lower jaw to the base of the tail; second, twelve lateral lights arranged irregularly along the head and body, and third, a series of four to eight

median light scales, either above or below the base
of the tail.

From the very first I directed all my attention
to the possible utility of these lights. The lower
battery, when going full, cast a solid sheet of light
downward, so strong that the individual organs
could not be detected. Five separate times when
I got fish quiet and wonted to a large aquarium, I
saw good-sized copepods and other creatures come
within range of the ventral light, then turn and
swim close to the fish, whereupon the fish twisted
around and seized several of the small beings.
Once it turned completely on its back. I could
never have seen this except that the glass sides of
the aquarium reflected sufficient light. Whether
this is the chief object of the ventral lighting I do
not know, but it is at least occasionally effective.

Perhaps the best distinction between various
species of this group of lantern fish is the arrangement of the lateral light spots,—indeed in the
dark-room I could tell at a glance how many
species were represented in my catch by their luminous hieroglyphics. When several fish were swimming about, these side port-holes were almost
always alight, and thus it seems reasonable to suppose that they are recognition signs, enabling
members of a school to keep together, and to show
stray individuals the way to safety.

The light scales of the tail are apparently of
great importance. Ordinarily when the whole fish
is glowing with the pale, cold, greenish light of
luminescence, these caudal lights are seldom seen.

A clue to their use is to be found in the fact that they show a remarkable sexual difference, the males having them on the upper side of the tail base, and the females on the lower side. Of course in my necessarily brief and sporadic researches, when no fish lived longer than thirty-six hours, there was no chance to observe courtship or any such use which these lights might subserve. But when a fish exerted itself unduly to get out of the way of another, either of its own or another species, these lights would flash and die in quick succession. Three separate times in unusually strong, vigorous fish when the body luminescence was very dim, these scale search-lights flashed like heliographs, being much stronger than the combined, steadier glow of all the others. This luminescence was of a much deeper green than that of the ventral lights. If continuously alight, a single fish would enable one easily to read fine print.

In the dark it was thus possible to distinguish species of lantern fish by the lateral hieroglyphic heliographs and the sexes by the upward or downward direction of the tail lights. I have never seen the latter illumination given out by a fish swimming alone in an aquarium. Although it is very evident that the caudal flashes have some sexual significance yet another very important function seems that of obliteration. It certainly was to my eyes, and I have no reason to think that a natatory enemy might not also be frustrated. When the ventral lights die out they do so gradually, so that the eye holds the image of the fish for a time after

their disappearance, but the eye is so blinded by
the sudden flare of the tail lights that when they
are as instantly quenched, there follow several seconds when our retina can make no use of the faint
diffused light remaining, but becomes quite blinded.
A better method of defence and escape would be
difficult to imagine.   Although I sometimes captured twelve hundred lantern fish in a single hour's
surface haul, the wonder of this animal illumination never became less marvellous.

An hour or two after the first Myctophids had
come to the surface, I would occasionally find a
somewhat larger, black fish among them.   In the
glare of the laboratory electric lights this was not
a very unusual-appearing fish, although it had a
short, dependent chin tentacle and a mouth with
exceedingly wide gape.   It was a fish named
Astronesthes (Col. Plate V), and for once the
ichthyological Adam had showed imagination, for
these Greek words mean "An eater of stars."   Not
until I dissected one did I realize the full significance of this title, for in each *Astronesthes* I found
a full-sized, just-swallowed lantern fish, although
the former was only about one-third longer than
its prey.   In the dark, this voracious black fellow
was a gorgeous sight, the skin covered with a host
of minute luminous specks, while the fins fairly
glowed with pale green light.   Curiously enough,
it was the stem, not the specialized tip of the
chin tentacle, which was luminescent.

But I have given more than enough space to
such plain unvarnished facts concerning these com-

mon fish of the nocturnal ocean.   On another trip, with a foreknowledge of the ease of examination of living specimens, and consequently a wholly new set of apparatus, I hope to approach much closer to the meaning of their lives.

CHAPTER IX

COCOS—THE ISLE OF PIRATES

BY WILLIAM BEEBE AND RUTH ROSE

I LOVE to think of the meeting places of the great elements, as where I sit in my bow pulpit in mid-Pacific. The sea is mirror calm with only the silent slipping past of lazy swells, more like evanescent breaths on glass than actual movement. So clear and blue and still is the surface that I cannot tell where the liquid begins and the air ends. Now and then the bow dips and my feet gently sink below the surface. The air is quiet and neither hot nor cold, and the world is perfect, with all mankind and his works out of sight behind me. I sit and solemnly make notes on the creatures I see, I ponder and wonder, and finally I am utterly discouraged at the thought of hoping to know the things of this planet any more clearly. Then comes a comforting thought, that after all I cannot expect to do much with a brain which has only one-ninth of tissue and substance to hold together the eight-ninths of water.

Five distinct and separate smudges of rain beaded the horizon, and as my eye played idly over these, one cloud lifted with amazing rapidity and revealed Cocos Island, clear and green, as the handkerchief of a conjurer is raised and displays a bouquet of exquisite flowers where a moment before there had been nothing.

I climbed at once up to the bow where I commanded a wider prospect. As Cocos—alive with legends of pirate hoards of gold, with every headland and inlet named after some brigand of the sea—as Cocos appeared before us, our approach to the island became perfect,—our escort began to form. As we neared it, great numbers of dolphins, those souls of drowned sailors, raced toward us in tens and tens and twenties, and gathered in all but solid layers about the bow and along the sides. I have never seen such hosts packed together. When we slowed up so that we could photograph them to better advantage, they all slackened speed and merely dipped and curved lazily in one spot, sighing as they exhaled.

Long before the island showed any detail, boobies, the long-familiar red-foots, and a wholly new green-foot, hailed us as the newest things in convenient perching places, the best dead trees they had ever seen, and our ratlines and wireless were crowded so that the birds touched each other. A few frigatebirds passed, some pure white terns swooped in the distance and—Cocos vanished. Over it, dark clouds materialized out of nothing, and the smoothness of the forested mountains became blurred and streaked with rain. Then a great curved arch of pale grey etched into the black rain cloud, a stain of some indefinable color appeared, deepened, and the island was crowned by a rainbow so brilliant that its edges seemed to carry human vision much farther along the scale than usual,—our eyesight almost interdigitating with heat on the one side and sound on the other.

With the disappearing of Cocos into the enveloping rain-clouds, my mind went back into equally obscure years of past centuries, when mankind first sighted this oceanic speck.

The discoverers of the Galápagos did not think enough of their find to attach a name to those islands which they were the first to see. But the man who discovered Cocos was even more indifferent, for he appears not to have so much as mentioned the circumstance that he had chanced upon this scrap of tropical jungle afloat in the Pacific, far from sight of any other land. At least, such a conclusion is an explanation for the fact that there seems to be no record of the first voyager to set eyes on its steep shores, laced with waterfalls.

The first map on which Cocos is shown is that of Nicholas Desliens, in 1541. This was six years after Berlanga, Bishop of Panama, reported his accidental discovery of the group afterwards known as Galápagos. No doubt Cocos was as fortuitously found, perhaps by some Spanish captain exploring the new domains of the mother-country, perhaps by a filibuster fleeing with booty from the mainland. Malpelo Island was already well-known, as most of the ships plying those waters could hardly have helped seeing it, and the Galá-

pagos were scattered over such a wide expanse of
ocean that they were likely to be seen now and
then.   But Cocos had an elusive quality which it
has not completely lost even yet.   Surrounded by
strong and tricky currents, concerning which much
remains to be discovered, and very often veiled by
such heavy mists and rainstorms that a ship may
pass within a few miles without glimpsing a trace
of land, the very existence of such an island has
been denied in comparatively modern times.

Of course in the 16th century, when navigation
was more an art than an exact science, it is easy to
understand that the precise position of Cocos was
difficult to establish, and long after its discovery it
was located at the caprice of the geographers, now
south of the Equator, now north, moving from side
to side, and on some maps completely ignored.   For
a while another island called Santa Cruz was fig-
ured as lying to the northeast of Cocos, probably
named by some navigator, who, obtaining a wrong
position, thought that he had found a new island,
which was really the ambulatory Cocos.

At last Cocos came out fresh and green from
her shroud of rain, and we slowed, sounding every
few yards, drifting nearer and nearer until the
heights of Nuez Island were well abeam to starboard, and Cocos itself loomed high over us.   At
the signal, in seventeen fathoms, the chain clanked
and jangled through the hawse hole and we swung
around head on to the stiff alongshore current.

Here at last, on the ides of May, we were close
to Cocos, three hundred miles off Costa Rica.   It

rose before us tiny and mountainous, only about three and a half miles across, with two peaks in sight, deep-seamed with ravines, one of which was almost twenty-eight hundred feet in height. Here and there off shore were a scattering of rocky islets, but, as we later found, the bottom of the sea dropped abruptly downward in all directions. No greater contrast could be imagined than between Cocos and the Galápagos—the one wet and green, the others dry and brown.

Night came quickly, dark, with swift scudding clouds and an occasional hint of subdued moonlight. Hoarse, disembodied cries drifted down through the night, and the restless waters of Chatham Bay lapped along our vessel, as jungle grass brushes against the sides of a smoothly moving elephant.

Dawn broke with the silent impetus of the tropics, and breakfast on this day lost all hint of a social rite, and became a hastily performed physiological necessity.

Our atavistic pirate threw his tiny Panama dugout and paddle overboard, dived after, baled it, crawled in, and sped shoreward, in the same spirit with which a pilgrim comes within sight of the Kaaba. No devotee ever climbed the seventy-two steps of St. Anne de Beaupré with more reverence than Don Dickerman, tumbled ashore by the breakers, crept up the pebbly beach.

I followed quickly and our little outboard motor vibrated rapidly across the bay. Great shadowy forms passed beneath, and now and then we had to snap the tiny propeller out of water as a giant

grouper made a rush for it.   Smooth white sand
alternated with coral skyscrapers and volcanic villages, fathoms beneath the clear water.   I did not
realize at that time that soon I would be walking
the streets of this submarine world, and making my
manners to their inhabitants.   At the head of the
bay series after series of three great rollers curved
and broke, so I chose the eastern side where the
surge struck obliquely against a line of mighty lava
boulders.   Rowing in stern first, I chose a moment
of equilibrium, and leaped out, bracing myself as a
waist-high surge swept past.   Guns, nets and
cameras were passed along and our first day on
Cocos began.

Wherever we went the way was barred by vegetation through which we had to force our way.
The only passable paths were up the center of the
rocky streams which leaped and swirled down from
the high interior.   Four-fifths of the island is on
end, with slopes so steep that the trees are set in
at most acute angles.   The rain which falls heavily
for many months of the year keeps the island as
saturated as a sponge, and the squashy yellow clay
and dripping vegetation seem seldom to become
even approximately dry.

I walked along shore beneath groves of giant
tree-ferns whose lacey foliage fretted the sky overhead.   Every now and then a silver column of water
would appear, falling from high up on the mountain, to spend itself in spray and a trickle over the
pebbly beach.   The sun came out and the whole
island glistened like a jewel with a myriad facets.

I came to a large stream and found a great boulder a few yards inland which gave a sight of the shore and of a glade at the forest's edge. Great orange and black brassolid butterflies hovered about the masses of morning-glories, hibiscus and clusia blossoms near by, while my view seaward was seamed by a hundred vertical lines of aerial rootlets, dropping from fig-trees high above.

A sharp cry drew my attention to a bird swinging in a curve out from shore, sandpiper fashion, and when it alighted I knew it for a wandering tattler. Then a black spot on the sand exposed by the ebbing tide turned out to be a grey Galápagos gull, so interesting a straggler that I later secured it. It was pecking at an old fish, and as I watched I saw a small something run a few feet away. My glasses showed a large rat—apparently of the usual ship's kind—mangy to an unpleasant degree, much of the hair being gone from its back. It was munching a bit of old fish. Cocos was revealing strange inhabitants with still more strange habits.

Suddenly the island, and tree-ferns and tropical-smelling jungle vanished in the haze of memory which a happy, lilting little song aroused, and on a branch a few feet away a yellow warbler sat and sang to me over and over his simple lay, which so often has meant early spring in my northern home. This bird is the same as the Galápagos warbler.

There are only four species of land birds on Cocos and later in the morning, within a period of fifteen minutes, I saw all of them without moving from my boulder. A flash of rufous and a throaty note revealed the only species of insular cuckoo, my warblers were all around me, and then there came to my ears the sharp snap of a bird's beak and on the tip of the barrels of my gun which I had left propped against a rock, perched the Cocos flycatcher, hardly to be distinguished from the little olive-green Galápagos chap.   In silence, finally came a small flock of the only finch, anomalous little birds with rather slender curved beaks, the males in black, the females mottled with olive and buff as though permanently saturated by the everlasting rain.   They flitted from twig to twig, playing at warblers, finches and titmice in their feeding habits. All the species of birds were seeking flying ants, small beetles and caterpillars.

A favorite feeding ground was at the limit of high tide where I saw all but the cuckoo again and again.   Here, too, came the ugly rats and twice we saw domestic cats, quite as wild as leopards, tearing at decayed fish, snarling at us and dashing away at our approach.   The birds were as tame as those of the Galápagos, and when they were not seeking for food they were investigating us.   On almost every tree were little Anolis lizards, scampering up and down the bark, and in flecks of sunlight expanding their relatively enormous, flat, bright yellow throat wattles both to charm their mates and to intimidate their rivals.

I picked out the nearest ridge summit and struck upward along an open grassy slope which, from the *Arcturus,* had looked like soft clover.   In reality it was far different—a sort of elephant grass, six to eight feet in height, with a saw-toothed edge which would cut to the bone if rubbed the wrong way. This we proceeded frequently to do, and when half-way up and making our way on knees and elbows, we discovered a species of nettle hidden here and there, and this was varied by an occasional nest of stinging ants. When we reached the summit I decided to return by a circuitous route through a deep, jungle-filled gorge.

Here we had only to slop and slither through the ferns and mud, now and then disentangling a rope-like liana which threatened to handcuff or garrot us as we descended. Being thoroughly drenched already and very warm we purposely fell into the first big pool of the stream, lay on our backs and commented with vigor on the delights of any extensive search after treasure in this difficult isle. Overhead we watched most curious sights. Here were hundred-foot trees growing so densely that the sunlight was dimmed to twilight, and high up on the topmost branches were perched scores of sea-birds—frigatebirds, boobies and pure white fairy terns—as out of place to our eyes as would be a cloud of dust in this saturated world.

On our way down we spread a small net across narrow reaches of the torrent and caught great crayfish and curious little vacuum-cupped gobies. Once we saw a giant a foot long, and on another day captured it.

I was astonished at the abundance of insect life, for other explorers of Cocos unite in dwelling on its scarcity. We took moths, large and small, including two species of beautiful, pink-spotted sphinx, several kinds of butterflies, large brown-winged grasshoppers with enormously long antennæ, funny little green cicadas, ants, mosquitoes, one small wasp, wood roaches large and small, and many giant dragonflies. Almost all these insects were clad in dull shades of black and brown, as were numerous beetles—elaters, long-horned, and weevils. One startling exception to this coloration was a weevil which stood out from the rest of the living creatures of Cocos as the daily rainbow contrasted with the somber storm clouds; indeed this tiny gem had its wing-cases dusted with a powder so glorious that under the lens it gave back every color of the spectrum, with emerald green as the dominant tint.

On my return to the *Arcturus* I frightened up a quartet of yellow-legs which flew after the tattler, and high overhead a hawk circled, the only one ever recorded for this island. The motor boat was anchored out beyond the surf and after fighting my way through the breakers and reaching the bow I saw a small green heron rise from the stern.

Directly after lunch I dived a number of times near the western side of the bay, at the first plunge taking down a good-sized aquarium with me. I had found it quite impossible to harpoon or catch the small blennies and other fish which crept about close to the coral, and many of which were new to me. In spite of the heavy surge I balanced the aquarium on its side on a block of lava, and baited it with several limpets, then waited, half floating in mid-water, with a sheet of glass in my hand. When two rare, rose-colored blennies had entered I slipped the glass across the opening and had accomplished my object. But then things began to happen,—a heavy surge washed me sideways and not until I saw a mob of little fish excitedly clustered about my fingers did I notice the pale red cloud in the water and realize that I had cut my hand on the glass, and that the blood had drawn a swarm of these dainty vampires. Then a series of strenuous jerks on the hose notified me that something had gone wrong in the upper world, and I left the aquarium and began to grope and leap toward the ladder. To my surprise I seemed to make little headway, and then I looked back and saw that the anchor rope had chafed through and that my slender length of hose was the only connection between me and the boat. I redoubled my efforts and soon the Jacob's ladder caught for a moment on a projecting coral ledge and I reached it and swarmed up. A few days later we made an effort to find the aquarium but could never locate it. However, I know that the glass trap will work and I shall try it again some day on a much more extensive scale.

It was as well that I came up when I did, for by the time I reached the *Arcturus* a wicked blue-black squall was headed for us, and the waves were too high and choppy to be safe for a small boat. The storm came from across the island. I called back all the boats and made sure everything was safely tied down. The rain advanced in sharply delineated clouds, and instead of a solid sheet of flat

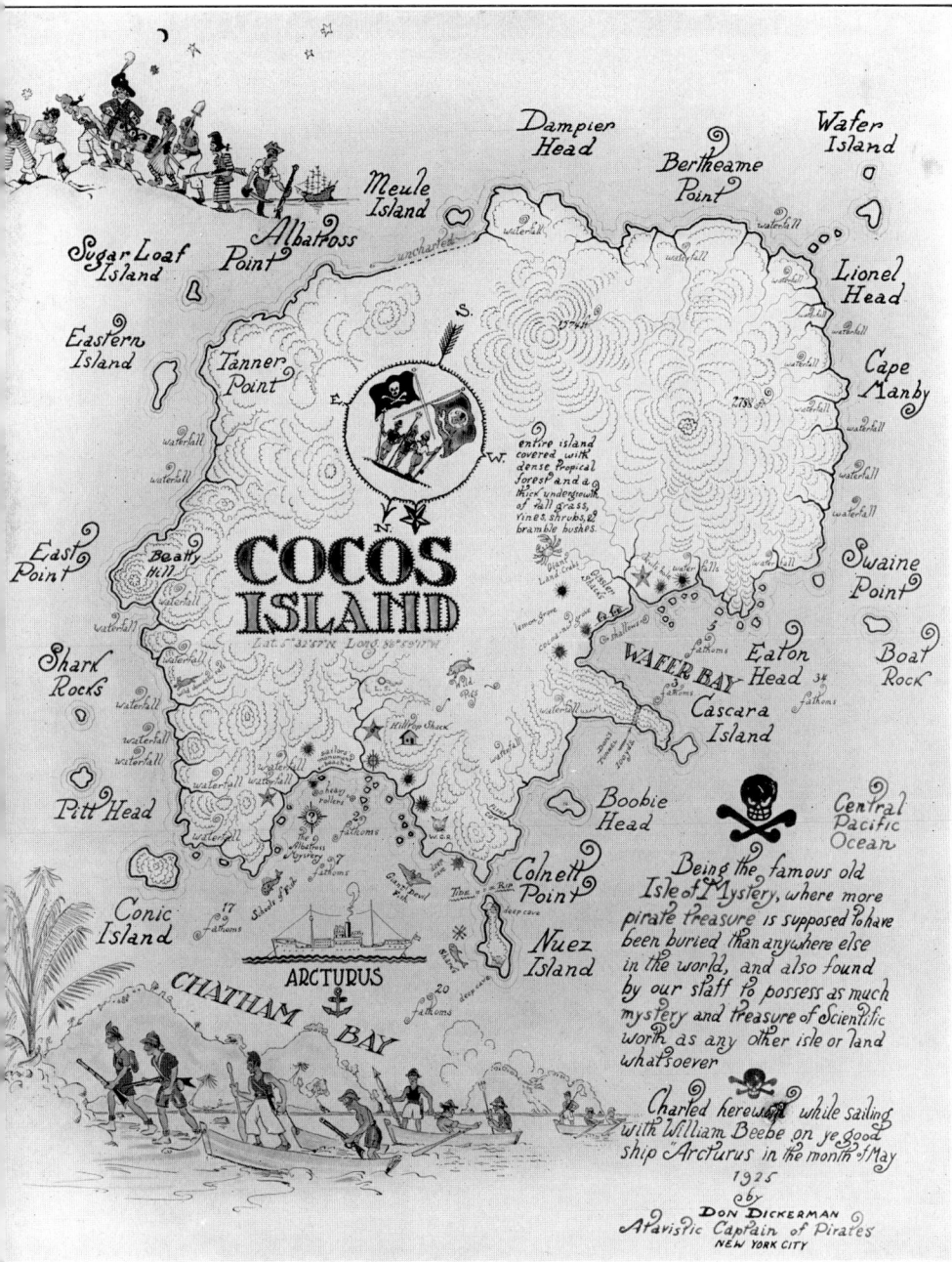

FIG. 36.—UNSCIENTIFIC MAP OF COCOS ISLAND.

Fig. 37.—Chatham Bay, Cocos Island.

Halfway down the beach is the fresh-water stream where pirates and whalers for hundreds of years have quenched their

green the island showed a host of unsuspected ravines and peaks, the onrushing downpour filling the
first and silhouetting the latter, giving depth and
perspective to the whole island, before blotting it out.

About nine o'clock at night the wind arose in
earnest and was soon blowing half a gale. All the
boats were tied alongside, rocking and pitching in
the rising waves. The night was black as ink, with
occasional squalls of rain, each followed by an
equally brief and even more sinister duration of
calm. Our lights were all on and the brilliant glow
made the surrounding darkness the more impenetrable.

Throughout this entire night of storm, boobies
by the hundred and noddy terns by the score flocked
to the steamer, covering the deck and filling the
boats. They seemed to lose all control of themselves
when they came within range of the glare from the
electric lights. Yet they did not dash into the light,
but merely alighted near it and remained quiet,
or flopped about and fought with each other.
Nothing showed the complete absence of man from
this island as much as this. The terrific wind and
blinding rain utterly confused the birds. All doors
had to be closed, for otherwise they filled the state-
rooms and laboratory, and their long, thrashing
wings worked havoc until we ousted them.

Taking a bird by the tip of one wing I would
swing it about my head and cast it far into outer
darkness, when, like a boomerang, it would right
itself, describe a wide circle and return. I tied
my handkerchief to the leg of one giant, green-

footed booby, hurled it forth, and a minute later it was again at my feet, and I retrieved my property.

The Battle of the Boobies will never be forgotten by any of us. The silken swish of wind-driven rain, the thud and shriek of newly-arrived birds, the thrashing of powerful wings as they flapped against the deck-houses or engaged in gladiatorial combats, their hisses and screams when approached by us, and the shouts and helpless laughter of the embattled scientists, would have made a phonograph record that no uninitiated listener could have explained.

These birds nested by the hundred in low trees along the shore of Cocos, and were now returning from their fishing excursions to relieve their mates. Their crops were full of recently swallowed fish, and their first instinct after landing on the *Arcturus* was to deposit six to twelve neatly aligned, perfectly fresh fish on the deck. Comedy was added to this performance by the sight of Dr. Gregory, armed with a big enamel tray, solemnly following the waddling birds about, picking up, with a forceps, specimens of rare fish which the unfortunate birds had intended as breakfast for their nestlings. The food was chiefly small flyingfish, half-beaks and squids, with a scattering of smaller species, especially *Ophioblenny*. An ichthyologist never questions the source of his specimens!

Later in the night cross currents of wind set in and the small boats began to labor at their moorings, twisting on their painters and piling up on one

another. It became necessary to cut them adrift
and in the terrific sea to row them astern and hoist
them aboard. The sight was a strange one. As the
first sailor went down the Jacob's ladder he had to
kick one or two birds off each rung, and the boats,
both inside and along the gunwale, were a perfect
paste of boobies. It reminded me of old Japanese
prints of boats filled with tame fishing cormorants.
As we looked over the side, the air was filled with
hundreds of squawking birds as large as geese,
dashing through the spin-drift and the foaming
crests of the waves. The sailors tried to protect
their faces from the flying birds and at the same
time to manipulate the half-filled boats. It was a
risky, cleverly-executed piece of work, and ultimately all were saved (Fig. 40). Our memory was
thus enriched by another unexpected experience.

Next morning the sun rose in a blaze of golden
copper—a third of the sky being molten, the rest
cold blue, while to the west beyond the island was
the inevitable rainbow with its end buried deep in
some inland ravine. Almost at once the sun was
quenched and there was only the oily rolling sea
covered with dead or sodden birds, and arched by
the equally sodden sky.

The waterfalls on the mountain sides had increased to foaming torrents, and tide lines were
conspicuously marked with floating tree-trunks,
branches and vines, and millions of green leaves,
extending in a straight line along the axis of Chatham Bay and on to the northern horizon. With

the coming of daylight we saw that our big *Alba-tross* launch, which had been anchored far inshore, had worked loose and vanished, and a search of twenty miles out to sea failed to reveal her. From the fo'c'sle there came excited rumors of the launch's engine having been heard during the storm of the past night, and hard on the heels of this myth were ready explanations of strange men who had found treasure on shore and, until now, had hidden from us until opportunity offered for escaping with their loot. But the prosaic fact was doubtless that she had sunk after chafing through her anchor rope, and is now the home of countless fish and octopus. In her is one of my diving helmets, a pair of Zeiss glasses, a bathing suit and a box of cartridges—all of which are now returning gradually to their original elements far beneath the restless waters of Cocos.

Man has had so little to do with this speck of an island that all the historical facts we can gather are of interest.

As early as 1600 the Dutch circumnavigator, Oliver de Noort, tried to find Cocos and failed, and in 1615 another Dutchman, George Van Spilberg, wishing to get coconuts and water there, missed it because he had its position as south of the Line.

In 1684 Dampier had a similar experience aboard the *Batchelors' Delight,* so we lack the careful and accurate account which this conscientious observer would no doubt have written. Instead he quotes hearsay:

"The Island Cocos is so named by the Spaniards,

because there are abundance of Coco-nut Trees
growing on it. They are not only in one or two
places, but grow in great Groves, all round the
Island by the Sea. This is an uninhabited Island,
it is 7 or 8 leagues round, and pretty high in the
middle, where it is destitute of Trees, but looks very
green and pleasant, with an Herb called by the
Spaniards, Gramadael. It is low Land by the Sea-
side.

"This Island is in 5d. 15 m. North of the Equator; it is environed with rocks, which makes it almost inaccessible; only at the N.E. end there is a
small harbour where ships may safely enter and
ride secure. In this Harbour there is a fine Brook
of fresh Water running into the Sea. This is the
account that the Spaniards give of it, and I had
the same also from Captain Eaton, who was there
afterward."

The *Batchelors' Delight* was then on the way to
attempt Realeja, and at that place Dampier chose
to go with Capt. Swan in the *Cygnet* to the East
Indies. Under Capt. Davis, on the *Batchelors'
Delight,* was Lionel Wafer, "Chyrugeon." Soon
after leaving Realeja, the crew began to fall sick,
so Davis put into the Gulf of Amapalla. They lay
there several weeks during which time 130 men
came down with spotted fever, and many died.
Wafer says, "Our men being tolerably well recovered, we stood away to the South, and came to
the Island Cocos. 'Tis thick set with Coco-nut
Trees, which flourish here very finely, it being a
rich and fruitful Soil. They grow also on the skirts

of the Hilly Ground in the middle of the Isle, and scattering in spots upon the sides of it, very pleasantly. But that which contributes most to the Pleasure of the Place is, that a great many Springs of clear and sweet Water rising to the top of the Hill, are there gathered as in a deep large Bason or Pond, the Top subsiding inwards quite round; and the Water having by this means no Channel whereby to flow along, as in a Brook or River, it overflows the Verge of its Bason in several Places, and runs trickling down in many pretty Streams. In some places of its overflowing, the rocky Sides of the Hill being more than perpendicular, and hanging over the Plain beneath, the water pours down in a Cataract, as out of a Bucket, so as to leave a Space dry under the Spout, and form a kind of Arch of Water; which together with the advantage of the Prospect, the near adjoining Coco-nut Trees, and the freshness which the falling Water gives the Air in this hot Climate, makes it a very charming Place, and delightful to several of the Senses at once.

"Nor did we spare the Coco-nuts, eating what we would, and drinking the Milk, and carrying several Hundreds of them on board. Some or other of our Men went ashore every Day. And one Day among the rest, being minded to make themselves very merry, they went ashore and cut down a great many Coco-trees; from which they gather'd the Fruit, and drew about twenty Gallons of the Milk. Then they all sat down and drank Healths to the King, Queen, etc. They drank an excessive quantity; yet it did not end in Drunkenness: But how-

ever, that sort of Liquor had so chill'd and be-
numb'd their Nerves, that they could neither go
nor stand: Nor could they return on board the
Ship, without the Help of those who had not been
Partakers in the Frolick: Nor did they recover
from it under four or five Days time."

Clipperton was the next famous person to visit
Cocos. He was there in December, 1720, and
January, 1721, and the record of the voyage is
quoted from a book by one William Betagh, captain of Marines with Clipperton's expedition, which
was "chiefly to cruise on the Spaniards in the Great
South Ocean." This book seems to have been written for the special purpose of confounding Captain Shelvocke, who was in command of the *Speed-
well,* under Clipperton on the *Success.* Betagh is
full of grievances which burst out on every page,
the very first being that Shelvocke did not appear
at the rendezvous at the Canaries, so that Clipperton had to sail without any of the stores of wine
and brandy which he had expected to tranship from
Shelvocke's supply. "And I own it was very hard
to be forc'd on a long voyage to the southward,
when the sun was in his northern course, without
either of those chearful supports of nature."

In January, 1720, Clipperton scrubbed ship at
James Island, and a fortnight later captured a ship
which carried the Marquis de Villa Roche, President of Panama, his wife and child. The Marquis
was an old acquaintance of Clipperton, as the latter, captured in former years in these waters, had
been taken before the Spanish dignitary. Now,

with the tables turned, the Marquis was held for ransom, but his wife and child were put ashore.

All through that year the *Success* cruised about, having various adventures off the coast, none of which came to much. A plot to seize the ship was discovered among the crew, as they were discouraged over the apparently fruitless voyage. At last, on Dec. 6th, "being unwilling to lose more time, we make our best way to the isle of Cocos, where we hope certainly to get fish, fowl and cocoanuts; our people being very sick and weak.......
.......17th.......at nine forenoon with joy we beheld the island Cocos about nine leagues N.W.

"18th. Anchor in 13 fathom white sand. Here all our people and the Marquis de Villa Roche got ashore, where we build a house for the sick men. Here is abundance of good fish round the island which we take pains to catch, the surf being sometimes very great. Our people find here plenty of coco-nuts, crabs, boobies and their eggs, this being their hatching time. Our captain broaches the last hogshead of brandy, allowing every man a dram a day; and on new-year's-day gave the people a gallon of strong beer for six. This food, ease, and refreshment pretty well recovered all our company. We wood and water, tho with much difficulty; for here is a great swell coming in from the northward constantly at full moon and change, therefore are forced to wait till the spring tides are abated before we can get anything off.

"Jan. 17, 1721. The Marquis came aboard as do most of our people, being ready to sail. Eight negroes and three of our men desert here and abscond in the woods. The names of our men are Higgins, Caulker and Shingle. The anchorage here being rocky we have sadly gaul'd both our cables. After continuing here a month, we weigh and set sail, from whence I take my departure, January 20th."

It would be interesting to know the fate of the stout Anglo-Saxon, aptly-named trio, Higgins, Caulker and Shingle.

In 1740 Anson sighted the island but did not attempt to land, altho his ships needed water. A scepticism which was unfortunate, in this case at least, kept them from benefitting by the lavish supply of precious water, and is voiced by Richard Walter in his record of Anson's voyage: "Indeed there was a small island called Cocos, which was less out of our way than Quibo, where some of the Buccaneers have pretended they found water; but none of our prisoners knew anything of it and it was thought too hazardous to risque the safety of the squadron and expose ourselves to the hazard of not meeting with water when we came there on the mere authority of these legendary writers of whose misrepresentations and falsities we had almost daily experience."

In the summer of 1793 Colnett was here on the *Rattler,* on a voyage which had for its purpose the extension of the spermaceti whale fisheries, and the investigation of anchorages which would be useful to the whaling fleets. He had come from his first visit to the Galápagos, where he christened Hood

and Chatham Islands, when he stopped at Cocos,
and from thence he went to the coast of Mexico,
and the Islands of Socorro, Santo Berto, and
Rocka Partido. He seems to have been almost as
devoted to his crew as the celebrated Captain Reece,
commander of the *Mantelpiece,* and in his account
of the voyage there is a careful description of his
methods of keeping the men in good health.

Captain Colnett has left us a long, accurate account of his visit to Cocos, much of which has to do
with helpful suggestions for future visitors. He
suffered almost continual rain during his stay at
the island, and concludes:

"We were much wearied, during the four days,
we passed off this island, and prepared to quit it.
We therefore took on board, two thousand cocoanuts; and, in return, left on shore, in the North
Bay, a boar, and sow, with a male and female goat.
In the other bay, we sowed garden seeds, of every
kind, for the benefit and comfort of those who
might come after us. I also left a bottle tied to a
tree, containing a letter. Over it I ordered a board
with a suitable inscription, which Captain Vancouver thought proper to remove, when he anchored
at this isle, some time after me. The letter gave
only an account of my arrival and departure. Having made the necessary arrangements, we set sail
for the Northward."

Colnett and Vancouver were shipmates under
Cook on H.M.S. *Resolution* in 1772-1775, Colnett
as midshipman, Vancouver as able seaman. But
when, in January, 1795, Vancouver, returning from

the Sandwich Islands with the ships *Discovery* and *Chatham,* put in at Cocos for water and found the letter left by Colnett two years before, he betrays no sign of ever having heard of the man before. His objections to recognizing this as the Cocos described by Dampier and Wafer seem incomprehensible, as there is nothing extant in their accounts which appears radically or insuperably different from the reality. As for his remark that "this island cannot be considered as having a pleasant appearance in any one point of view," and his references to the "dreary prospect," one can only accuse him of having the Englishman's occasional affliction in the tropics, a "liver" that casts a pathological gloom over the fairest landscape.

Chatham Bay was named for Vancouver's armed tender, which lay at anchor there.

It is to Sir Edward Belcher in His Majesty's Ship *Sulphur,* that we owe the survey of the coast of Cocos which is the basis for most of the charts of the island. Although Belcher made his maps in 1838, there is to this day an incomplete portion on the south marked "uncharted," a striking illustration of how little interest has been taken in this isolated spot,—at least from any point but that of treasure-seeking.

It is worthy of note that although Belcher was here almost twenty years after the supposed date of the burial of the first lot of treasure, he makes no reference to it, and it is fairly safe to assume that he had not heard of the hoard. Probably the fame of Cocos, until the time of Keating's discovery of treasure in 1845, rested wholly on the ease with which unlimited quantities of sweet water could be obtained, and during the years of the whaling industry, it was a resort of ships, a place of rest and refreshment, a place to explore, in the intervals of getting wood, water and coconuts aboard. Certainly they did not spare the coconuts and with characteristic indifference to the future, they obtained the nuts by the easiest method, which was to cut down the trees. There are still coconut palms on the island, but pitifully few compared with the old descriptions in which they seem to have been almost the dominant vegetation. During Captain Gissler's tenancy he planted more, as well as plantains, limes, coffee, and various vegetables.

Captain Belcher writes, "On the 3rd of April, 1838, we made the island of Cocos and on the following morning observed two whale ships at anchor. . . . On landing, I was surprised to find a hut and several seamen, one Portuguese, one English, and five blacks, Americans, landed by their own demand from one of the American whalers. At first I suspected foul play, but on the masters of the vessels landing and stating the facts to me in the presence of the men, they acknowledged that they preferred living on the island to sailing in his vessel. Their contract was only 'from the Sandwich Islands until they reached a port.' They were evidently bad characters. Their only subsistence was fish, pigs, boobies, noddies and other marine birds frequenting the island.

"Water is very abundant and was easily conveyed by hoses into the boats. . . .

"In Chatham Bay we noticed the rock mentioned by Vancouver, and left on another the *Sulphur's* name. . . .

"Fish are abundant in Chatham Bay, but were not easily taken at the ship. The whalers sent their boats daily to fish in the tide stream between the small island and the main, and were very successful. Shell fish were scarce and few worth preserving. . . .

"It was not without surprise that I read Vancouver's opinion of this island. The view of the two bays, with the magnificent S.W. cliffs and waterfalls, like silver threads, leaping from the richest and varied tints of green that can be imagined, would put a painter in ecstasy. Season, however, may make a material difference. The same objects we view and are delighted with in sunshine, are dreary and uninteresting in gloomy weather.

". . . The thicket is not now impenetrable, as the self-exiled whalers traversed easily from bay to bay. Goats are said to abound but keep to the heights. Pigs are plentiful and one large hog was sufficiently inquisitive to look into the tent at a distance of twenty yards.

"The stream in West Bay produces fresh-water fish but we could not obtain any. A curious bullhead was taken, as well as fresh-water crustacean, at our watering-place. Some of our men, who had landed to wash and amuse themselves, found their way up the hill east of the water-course, and saw into the interior, which they described as a lake, or large sheet of water. This would account for fresh-water fish in West Bay. The quantity of water we had noticed in streams, waterfalls, etc., and which were not much augmented by heavy rains, or by the stream in our immediate vicinity, must be supplied from this lake. No rains could preserve the volume and equality for twenty-four hours."

He also planted vegetable seeds.

"Before my departure I used every persuasion with the masters of the Americans to take these unfortunate people away, as well as pointing out to the people themselves the misery they must endure, and the foul suspicions which the next vessel would entertain of their conduct; but only one embarked."

A year later, April 7th, 1839, "at nine we anchored, all heartily anxious to escape a rainy season in our present jaded state. An American whaler, according to their praise-worthy habit of assisting any friend in view, sent her boats to assist in towing the *Starling* to her anchorage; but we were too far out to partake of her aid.

"On the morning following, I landed to obtain observations, and the early part of the day certainly led me to anticipate all I looked for, but noon destroyed my hopes, the rain falling in a complete deluge. I succeeded, however, in obtaining the requisite data, and also witnessed the effect of the heavy rains on the streams; converting a very quiet brook into a turbulent rapid in the course of a very few hours.

"On my last visit, I mentioned that three men

were left behind by an American whaler. These had remained a considerable time on the island, but were eventually taken off by another whaler; not, however, without poisoning the minds of part of her crew, two of whom were induced to try a similar experiment, and were now almost reduced to starvation, notwithstanding the presence of their countrymen. The master, however, assured me of his intention of giving them a passage to Payta, the lesson of the former characters leading him to assume severity to the last moment, as a warning to his own, as well as to the crews of other vessels."

"The pumpkins had flourished, the whaler having collected fifty from seeds planted just a year before. The other seeds seemed to have been destroyed, but he planted more."

The more recent historical events connected with Cocos belong more especially to the following chapter, in which Miss Rose treats of the era of treasure hunting.

One of my last memories of Cocos is the most dramatic. I rowed across to the west side of Chatham Bay close to Nuez Island. This is a very lovely sliver of land, a few hundred yards long and with a steep, high ridge, the underlying rocks showing through the foliage—white shoulders through a tattered coat, while, on the lower reaches, long, flowing grass clings like the exquisite emerald pelage of some somnolent behemoth. The figs and other trees drop showers of aerial rootlets which drape the island like gigantic beaded curtains.

Every time I looked at Nuez from the *Arcturus* there came to mind Böcklin's Töten Insel.

I anchored the flat-bottomed diving boat fifty feet off shore in the quietest spot I could find and then submerged in about thirty feet. Visibility was remarkably good and I could see clearly for one hundred feet in every direction. On one side enormous boulders piled themselves up higher and higher until they crashed through into the air and on up the slopes of this isle of death. In other directions the bottom sloped gently but steadily downward until it was lost in mysterious blue depths toward the abysses of the sea.

The swell was heavy and the end of my swaying ladder reached alternately from twenty to within ten feet of the bottom, as the boat rose and fell on the surface. I knew my leaping ability in this gravitationless medium so I did not hesitate to drop from the lowest rung at the moment when it was nearest the coral floor. I landed on a table of lava and was at once the center of a school of great grazing fish, triggers, parrots and surgeons, the largest I had yet seen, with now and then a unit of swift carangids, gleaming like purplish jade as they shot past. Out of the blue distance there materialized a man's-length of white-finned shark, then another and another, until sixteen were milling slowly about between me and the surface. This was a new habit and an unexpected formation of these fish and I must admit that the ladder looked very long and very high above me. I was so uncertain of the significance of this gathering that for

FIG. 38.—THE *Arcturus*, FROM THE SHORE OF CHATHAM BAY, COCOS ISLAND.

Fig. 39.—Wafer Bay, Cocos, with Fresh-water Stream in the Foreground and Cocoanut Palm Planted by Captain Gissler

a time I crouched in a circular cavity between two great coral growths, with my helmet in the entrance like a cork in some astounding bottle. The sharks showed no more than curiosity and, as usual, I was much more concerned with the ugly four and five foot groupers who pushed their unpleasant mouths within a few inches of my body and limbs. But when I saw the pigfish and the angelfish swimming unconcernedly about, I took heart and strode forth.

In the dim distance I could see a very beautiful sea-fan and started for it. Never, even in the high Himalayas, have I ever breasted so stiff a wind as the push of this current which swept past Nuez. At times I was lifted clear off my feet and carried back. Twice I found myself at my starting point. So I went down on my knees, and with fingers and toes clung to every step which I gained. With me went the brobdingnagian groupers and the lesser fry of angelfish and always overhead circled the sharks. My hose had trailed behind now and was no longer the hub of their orbits. For a while I was the center of attraction in this part of the Pacific Ocean.

After much effort I reached my sea-fan and hung on to it while I floated in mid-water, waved about by the current like a rag on a bush. My body-guard had thinned out, and twisting around, I saw a tiger shark weaving slowly toward me. I would gladly have given my place to any eager scientist in the world, or relinquished it to one of the thousands of men with more courage than I possess. But at least I was not bothered with a choice of action—

there could be no thought of escape by flight. I crouched close to my wisp of sea-fan, although hiding behind it was as effective as an attempt to conceal oneself behind a handful of ostrich feathers.

The shark appeared enormous—thirty feet came to my mind. Then, like Dunsany's ghost-watcher at the Castle of Oneleigh, I sought to distract my fear with geometry. I estimated the shark's length, I compared it with other fish near it, and I was more composed by the time my mind settled on eighteen feet as the extreme length I could assign to it. I had already faced scores of sharks and even other tiger sharks in my diving, but never so large a one, nor in such unprotected surroundings nor without at least a grains in my hand. The great elasmobranch came on until I could see the black veins in its yellow, cat-like eyes, and the loose, adenoid-gape with its lining of triangular teeth. The mighty tail swept farther to one side, the shark veered—and passed. An unusually heavy surge once carried it far back toward me, but it never turned and soon vanished beyond the shadow of the boat.

When I breathed slowly again I braced myself and with all my might dragged the animal bush from its moorings, and waving it like a purple banner, I returned, leaning hard back against the current. With no take-off except a low crouch I leaped upward and slowly rose through eight or ten feet of water and seized the lowest rung. For a while I hung there, soothed by the lift and settling of the deeper curve of successive surges—surges which, fifty feet away, were crashing into foam against the cliffs of Nuez. I looked around at the scattering hosts of *Xesurus,* troggers and angelfish, watched a trio of sharks meandering casually along beyond them, and, with a final gasp of wonder at it all, for the last time in these waters I climbed until my upraised hand thrust the great purple sea-fan into the open air.

CHAPTER X

COCOS—A TALE OF TREASURE

BY RUTH ROSE

THE stairs creaked dismally, and an elevated train roared by, as I climbed the dimly lighted flights. On each landing a gas-jet wheezed and cast trembling shadows on closed doors, behind which muffled voices could be heard, and I wondered if any of those huddled lives would make as strange a story as that of the man I was going to see. My mind dwelt on things so incongruous to this setting as surf-pounded beaches, leaning palms, steep jungle hillsides, for I was in search of a sailor who for twenty years had dwelt with his wife on an otherwise uninhabited Pacific island, while he sought for pirates' loot that was buried in that lonely spot more than a century ago. Surely no one can dispute his title to the record for continuous, non-stop treasure-seeking. Beaten for the moment, but undiscouraged and with his faith unshaken, he is now living in New York, remembering tree-ferns under a tropical moon, while he watches the flicker of an electric sign behind the pillars of the "L."

August Gissler, German by birth, sailor by choice, first heard of the Treasure of Cocos in 1880. He has written a book on his wanderings in search of clues and his adventures on the island, which, when published, will give in greater detail than is possible here, the story of a lure that has drawn men —and women—from all walks of life, from a hundred different ports, that has made the name of Cocos known over the world, and that, in the case of Captain Gissler, was strong enough to hold him during a great part of his life. He has seen expeditions come and go during his sojourn, and is qualified to speak with authority of the effects on the human animal of those potent words "buried treasure." It is to his generosity that I am indebted for the use here of some of the material from his own book.

This chapter pretends to be no more than an approximation of the facts, for many of the tales of Cocos have more than one version, and it is hard to sift out the truth in those cases where completely trustworthy records do not exist,—or are, at any rate, undiscoverable; its chief interest must lie in the story of the man whose pluck and perseverance kept him to his self-appointed task in the face of every obstacle and discouragement.

Buried treasure! That is surely the most romantic phrase in the language, warranted to bring a sparkle to the dullest eye, and to quicken the pulse of a paralytic. Quite baseless rumors of such hoards have sufficed to cause excited stampedes, so it is not surprising that the tale of the cache on

Cocos, fairly well-authenticated as it is, should have inspired all sorts of people with a desire to try their luck, even when they had no more to go on than the location of the island, and a lucky feeling.

Captain Gissler tells one story of an expedition that started in Pittsburg, outfitted in San Francisco, and landed on his beach one day, filled with confidence and armed to the teeth. They were prepared for battle, murder and sudden death, treason, mutiny and marooning, but they had somewhat neglected the subject of just where the treasure was hidden. They had no tattered old map, no cryptic key; they had come from Pittsburg because "a man" had furnished them with the clue that the treasure was to be found a hundred paces from the wreck of a pirate ship. This upwards of a hundred years since a pirate had been in these waters.

Captain Gissler says, "When I showed them what fools they were, they could see it too, but they couldn't see it till I told them!"

Cocos looks very small on the map; when the would-be treasure-seeker finds that it is only three and one-half miles in diameter, (though as most of the island is on end, the actual distance covered in traversing it must be three or four times as much), his hopes rise, and he is seldom dashed by the additional calculation that this area comprises about sixteen square miles. But when he finds himself actually occupying a fraction of those miles, miles of dense thickets, close-woven vines, sharp-edged grass, steep ravines and all-too-frequent torrents of rain, the spade in his hand dwindles to an inadequate toy, and his compass-bearings seem always to bring him to spots where recent landslides have obliterated the clues that he was so sure of finding when he should reach this island El Dorado.

The origin of the treasure of Cocos goes back to the year 1820 or '21. At this time there was an ex-officer of the Portuguese navy, a man of good family, who fell upon evil days and turned pirate. His name was Benito, and he is generally referred to by the euphonious appellation of Benito Bonito. He lived up to the best traditions of swashbuckling and bloodthirsty piracy, and ravaged the West Indies and the east coast of South America with great success. He dealt savagely with his crew when necessary, and owing to certain prudent habits, such as never appearing on deck without a drawn cutlass in one hand and a cocked pistol in the other, he flourished for several years. At length the Caribbean became too hot even for one of his ebullient ways, and he rounded the Horn to ply his trade along the western coasts from Peru to Mexico. He made several rich hauls, and is supposed to have buried on Cocos a collection of loot worth millions of dollars. Not long after, he was captured, and he and most of his men were hanged. Two who escaped were called Thompson and Chapelle.

Some years later, after Peru had won her independence from Spain, that South American republic was in the throes of civil war and counterrevolutions. Much of the private wealth of Lima, together with quantities of church plate, was sent to the fort at Callao for safety. When an attack on the fort seemed imminent, and its impregnability doubtful, the treasure, estimated in the millions, was transferred to an English sloop that was at anchor in the harbor. The temptation of such riches between decks was too much for the captain and crew; during the night they killed the Peruvian guards, slipped anchor and stole away. The captain's name was Thompson, and he is said to be the same man who had sailed with Benito. A less fitting name for the ship in which this piratical deed was performed it would be hard to imagine; she bore the demure title of the *Mary Dear*. Whatever her nominal shortcomings, she was a fast sailer, and outdistanced pursuit. There was no port on the mainland for which they dared to steer, so when they came, by chance or intention, within sight of Cocos, it was determined to bury their loot till some more propitious time.

This accomplished, they ran for Central American shores, but were captured and taken into Panama Bay by a Peruvian ship. Here every man aboard the *Mary Dear* was hanged on the spot, except for Captain Thompson and one other. They were spared to show the hiding-place of the treasure, but they escaped by jumping overboard in the night, and swimming to an English whaler that lay at anchor. Here they concealed themselves until the whaler had been several days at sea, and on emerging, were welcomed by her captain as additions to the undermanned vessel. Thompson was evidently not born to be hung.

Thus Cocos became the Treasure Island, *par excellence,* of the world. Thompson, the god from the machine, the repository of secrets, the man who had twice seen tremendous wealth buried on the same small patch of land, is next heard of in 1844. In that year a ship bound from an English port to Newfoundland carried a few passengers, among them a man with an air of mystery. One of the sailors, a native of St. John's named Keating, made friends with him, and received the confidence that the passenger was not anxious to draw the attention of the authorities. On arrival, Keating took the man as a lodger in his own house, and observed that he never ventured abroad in daylight. At length the stranger revealed himself to his host as Thompson, told him of vast treasure on a Pacific island, and showed a rough chart of the Cocos depository. There are several versions of what happened next; one story says that Thompson died, another that he fled in the night from unknown enemies, still another that he went to London, where Keating followed and obtained further particulars. At any rate, Thompson, having served his turn, disappears from the stage for good.

The known facts are that Keating interested a firm of merchants in the project of recovering the treasure, and that they outfitted a vessel to go to Cocos. Keating was an illiterate man, unable to read or write, to say nothing of knowing anything of surveying, so one Captain Bogue was sent with him to figure out the positions given on Thompson's chart. There was also the ship's captain, Gauld, who considered himself the head of the expedition. The merchants had erred in so distributing authority, for there was bad blood between Gauld and Bogue long before they reached their destination. At first they quarrelled; then they sulked; and at last worked off their pent-up spite by leaving insulting notes to each other on the dining-table!

The most disastrous effect of this petty squabbling was that in the course of it the crew learned the object of the voyage, which had been a secret when they sailed. When at last they dropped anchor at Cocos, Keating and Bogue hastened ashore alone for a preliminary survey. Having (so Keating's story goes) verified the chart and located the treasure, they returned aboard with exultant looks, to be confronted by an openly mutinous crew with an ultimatum. They were all to share equally in whatever was found. In vain the two men protested that those who had financed the expedition were entitled to divide the profits. The crew became so threatening that Keating and Bogue, in fear of their lives, pretended to consent. That night, while the sailors were noisily celebrating the money that they thought was as good as in their pockets, the two leaders put food and water in the big whaleboat and cautiously pushed off from the ship. Stealthily they rowed to the beach, and in spite of darkness and dense undergrowth, located the treasure and brought to the shore as much as they could stagger under.

The rest of Keating's story is a mass of contradictions. He eventually reached St. John's, alone, where he exchanged gold pieces and some bars of bullion to the value of only about 1300 pounds. Bogue was never seen again. Keating said that he had been drowned, so laden down with gold ingots that he could not swim, but sometimes this tragedy was said to have taken place when trying to launch the boat on that exciting night at Cocos, and sometimes it happened in Panama Bay, which the two men were supposed to have reached in the open boat. There was a half-hearted attempt to try Keating for murder, but in the absence of a corpse, the case came to nothing.

Captain Gissler, who has weighed and sifted every available scrap of information, has a very plausible explanation of the two men's movements on the night of escape from the mutineers. He does not believe that Keating ever knew the spot where the treasure was concealed, but that Bogue left him to guard the boat (a very necessary precaution on Cocos beaches, with their heavy surf) while Bogue made his survey and paced off the distance. That was in the daytime, on their first landing. That night the same thing happened; Keating kept the boat from smashing on the beach, while Bogue groped his way to the cache, and returned with all he could carry. It might very well be that the two men disagreed then and there, and that the snow-white terns, roosting in the fringing trees, were wakened by angry voices and the sounds of a struggle. Or perhaps there was no noise but the

sound of one blow, and then the laboring breath of
the survivor as he strained and tugged to launch
the boat alone.

At any rate, Keating seems to have been a
marked man and to have led but a sorry life until
he died.   Those who did not shun him as a suspected murderer, fawned upon him as the posses-
sor of the key to vast wealth who might be flattered
into sharing his secret.   His second wife, years
younger than he, thought that in the course of
many curtain lectures she had surely obtained sufficiently detailed information to enable her to find
the remaining gold without difficulty, but as will
be told later, her search came to nothing when she
finally reached Cocos many years after Keating's
death.   In fact, what he told her led her to search on
a different side of the island from the place where
the treasure cave could be found, according to what
Keating confided to a man named Fitzgerald.

Casting back and picking up the trail of Chapelle, the second man of Benito's crew who escaped hanging, he left San Francisco in 1841,
bound for the South Seas, and was never heard
of again.   Before leaving, he turned over to a
friend some papers, among them an extract from
the log of Benito's ship, indicating the location of
the treasure.   This was the genesis of half a
dozen expeditions that at various times have fitted
out in San Francisco.

When in May, 1925, the *Arcturus* spent ten days
at anchor in Chatham Bay, we found abundant
evidences of treasure-hunters.   Rain-filled pits,

spades and picks perforated with the rust of years, corrugated remnants of huts, were to be found along the shores. The records of many ships were carved on great boulders at the mouth of the stream that rushes down the ravine and spreads in a broad shallow across the beach. The oldest ones might have been written in a strange invisible ink, for at low tide they were indecipherable on the sun-dried tablets of the rocks; then at high water, as the first breaker dashed over them, the weather-worn letters magically appeared like a picture resolving in a crystal. The earliest date that we could read was 1797, left by "HIS BRIT MAG' SCHr LES DEUX-AMIS." We looked in vain for Vancouver's name and date, to the carving of which he refers. Other inscriptions, some much garbled and incomprehensible, are here set down:

    Bk VIRGINIA—MARKS—   A. Savvely 1875
    H R VIGILANT Sept 3, 1862
    Jos GRANT   Nantucket
    ADDISON   1861
    VAPOR
    BARK, Tybee, Feb/58
    G. Duffy Oct. 30   1843
    BRIG Adeon   Wm Low Nov 20—1830
    J. Maria   ZELEDON   Julio 22   1879
    BK ANARORA   J M E
    P. H.   1851
    S. ENTERPRISE   N. T. Dec. 1855
    Ship SUSAN   P. Howland Aug. 26   1851
    BARK Java   Nov 14/56
    N. P.   J. R. Lawrence

P. Cleveland   1864

Schr. GEN. PIERCE

J. BOND, Marblehead

The GREENW — LEONID   Boston

GRANDᵂ HAMMOND

HENRY HALL of LONDON

CRETAN

FURY

SAAIL SHREW

LYS

MARIPOSAx 1:6 Px 1871 x 1870

C. MARKS   Sep. 1871

SHIP BLO

    A. A. Campau 1852

Ship JOHN

    Anderson

Ship Royal B. Oct. 1849   H. P.

JORUVE   Nov 1813

G. TODD   1813

WASHINGTON GARDNER 1807

T. NEWTON   Norfolk VA   M 1 3

H. B. M. STEAM FRIGATE SAMPSON 1847

      SIR C. SEYMOUR

          Cap. HENDERSON

SHIP INDIENCHIEF of NEW LONDON

    CPT. BALEY MARCH 28   1848

S. JOHN

BARKCORNELIA   P. B. ROLUFS—MST   April

   14.   1852

Brick des Mᵗᵉ Ie GENIE, Comᵐ PML Cᵗᵉ ᵈᵉ

                  GUEYDON

        I Nov. 1846

PF LIEN Sranville 1840

FRANCIS L. STEEL   Mar. 28.   1871

B. COLCOND   May 15   1863
TRIDENT   MAY 1863
S. H. HARRIS   N. Bedford   Apr. 15—1842
May 21   R. C. Fay   1842
G. N. MACY 1853
BARK BEN
SHIP UNCAS   H. C. Bunker   Falmouth   Sept.
   23   1833
Ship ALEXDR. COFFIN   D. Baker   Nantucket
   Oct 12 1833
SHIP ATALA   G. Winship   BOSTON   1837
BARK OCTAVIA   C. MANTOR   Suban Apr.
   18.   1831
W. U. DAVIDSON McH.   CAHU.   PAUOA.
   H U . M. 25
Ship KINGSTON   W. E. SHERMAN   1833
PETREL
HIS BRIT MAG' SCHʳ LES DEUX-AMIS
   Sept   1797
Sch ROSCOE   MAX—7-1870
C. H. DUNN
D. DACK
E. H. FISHER Jʳ   May. 7. 1863
JOHN HOWLAND
LOUISE
SROST   1855
T. O. Simpson   1850   Sept. 18
FERNANDO
BX HYDSPE   62

It was fascinating to attempt the reconstruction of scenes that have taken place here, and, like the detective of fiction, put ourselves in the places of those men who came here with untold wealth and were confronted with the problem of concealing it. There are only two possible anchorages, Chatham and Wafer Bay, neither of which permit a ship to lie nearer than a quarter mile from shore. We imagined the *Mary Dear* and her crew during the weary work of disposing of boatloads of specie and bullion. Anyone who has ever seen the place feels exhausted at the mere thought of their labors after the booty was landed. Of course they would not bury it on the beach, so they must have transported it painfully, a very little at a time, along the swift, rocky streams, or up the slippery hillsides and across the chasms with which the island is rent. And no one would envy them the task of excavating, in the root-filled, stony soil, a hole large enough to contain millions of dollars worth of precious metal.

On one of the two rainless days which we had during our stay, Betty and I set off inland. Having had experience of land routes, abounding in razor-edged grass, wet clay soil that converted the hillsides into tropical toboggan slides, stinging ants and, in spots, all but impenetrable undergrowth, we followed the stream bed from Chatham Bay. It is a wonderful river, both for its pictorial qualities, and for the unlimited possibilities of exercise that it affords. Where it leaves the jungle to flow across the beach, branches laden with blossoms arched across its gurgling cool shallows. Splashing along, for a while all our attention was concentrated on mere progress. Here we crossed a tiny sand-bar, with water only to our ankles; beyond would be a line of boulders, damming the stream so that a pool of swimming depth barred our way. We wanted to keep our collecting bags dry, but we had no such ambitions for ourselves; a week at Cocos would make anyone feel web-footed.

Then we would take to the bank, making a precarious way along the overgrown slope until the deepest places gave way to navigable shoals once more.

Not a hundred yards from the beach, is a huge boulder bearing a mark that has some resemblance to a sombrero; this has been regarded as a clue, and used as a landmark by many a hopeful explorer, and is generally referred to as "Benito's Hat." The stream was filled with stones, ranging from those that turned treacherously underfoot, to huge blocks six or eight feet high, over which we swarmed clumsily, clutching at any convexity and wriggling to the top, to descend the other side with a final splash. At a fork there was a tiny island, presenting in miniature a perfect environment for the collector. A rotten log, a stump, minute pebbly pools, small saplings, and a few large stones yielded to our inspection termites, ants, little crayfish, large locusts, and several species of spiders, as well as a beetle or two. The main branch of the river turned to the left, but as some of the party had already gone that way, we decided to try unexplored ground, and follow the right-hand, smaller stream. A bank of pebbles drew our attention to a cave in a rocky outjutting of cliff; it was a shallow depression under the overhanging hill, and in it were the bones of a large bird.

A few yards further on, a cascade fell sheer down from a sixty-foot height. There was no possibility of climbing up at either side, so we made a detour to the left, and searched for foot-hold. It looked fairly simple; there were plenty of trees to grasp and cling to, while the next step was considered, but we soon found those trees were rooted no deeper than grass. The tree-ferns were particularly deceptive; thick, sturdy-looking trunks grew out at convenient angles from the cliff, and the weight of a few pounds sufficed to dislodge them entirely, and with them a cartload of earth and rocks. Testing every cautious inch, with faces pressed against the ground most of the time, we wormed our painful way to the top without casualties. A sharp ascending ridge led us to a hill-top almost clear of undergrowth, where a few gigantic trees grew. They were as beautifully spaced as though a landscape gardener had planned the vista, and almost every inch of trunks and branches were covered thick with parasitic plants.

We stood panting, and watched four finches that flickered among nearby twigs. In the thick jungle tree-tops sat frigatebirds, looking as out of place as sailors on horseback. Lizards darted about in the bromeliads that encrusted the trees. The silence seemed unbroken and unbreakable. Then a tiny faint cry sent our glances upward. Not more than two feet above our heads hovered a snow-white fairy tern, dainty wings beating with incredible rapidity to hold it in this spot. Turning its head from side to side, it hung there for a long time, looking down at us, and we could feel the little breeze from those immaculate fanning wings. Standing in that lovely sea-girt forest, in the jungle hush of noon-day, we thought of a Biblical snow-white bird with no feeling of sacrilege.

Then another and another came, and the trio hovered fearlessly about us till, their curiosity satisfied, they fluttered away among the trees. We continued the wide circle we had commenced and finally rejoined the river, which at this height was no more than a small brook, flowing in a deep ravine, with sides so steep and slippery that it was impossible to leave the stream-bed, no matter what obstacles were encountered in it. The densely-wooded hills towered almost straight up at each side and made even midday gloomy to us at the bottom of this narrow crack. The vegetation was swamp-like in its luxuriance and character, and many of the plants were an unnaturally vivid green, like artificial things. When we reached the top of the waterfall that we had circumnavigated on the way up, we hung over the edge and watched the falling sparkle, while we chose a way down. Nothing could be worse than the way we had come up, so we decided to try the other side for our descent, and discovered that going down is always worse than climbing. I am sure that we defied gravitation most of the way, and the proceeding was not made easier by each of us finding the other's predicaments and postures irresistibly funny.

The last few feet we fell, abruptly and simultaneously, and landed near the little cave, caked with mud, thoroughly scratched, very hot and completely happy. With a piece of soap, brought with great foresight, we scrubbed ourselves and our garments in a deep pool, and floundered back to the beach, with a new respect for the buriers of Cocos treasure and the conviction that wherever they deposited it, it was not above that waterfall.

Three sailors from the *Arcturus,* in a cross-country scramble, found rusty shackles attached to a post that was deeply embedded in the ground, amid rotten boards that seemed to be the collapsed fragments of a hut. This gave rise to excited speculations, but later, in Panama, we found that the Costa Ricans (who own the island) had for a short time maintained a convict settlement on Cocos, of which these things were probably the relics. The prisoners must have heard of the golden store; perhaps the chance that one of them might find it lightened their penal labors.

Convicts, peers, philanthropists, journalists, middle-aged widows, sailors, adventurers,—they have all played their parts in the Cocos story. More than one company has been formed, has issued prospectuses and sold shares in proposed expeditions,—and more than one has never left port. The catalogue of ships that have actually sailed for Cocos would be a long and tiresome enumeration, but there are interesting details about some of them. In 1875 one of the Pacific Steam Navigation Co.'s ships came treasure-seeking, and one of the crew, Bob Flower, while scrambling over the island, slipped, rolled down the side of a steep ravine, and

literally fell into the treasure, or at least some portion of it. Tradition says that he brought away as
many coins as he could carry, and the story is
always quoted as being "well-substantiated," but
by what or whom it is not easy to discover. If he
found more than he could carry, it seems likely that
his shipmates would have been more than willing to
lend him a hand.

Two Englishwomen outfitted a ship and went to
Cocos, equipped for a stay of ten weeks. They announced that the treasure—when found—would be
used to establish an orphan asylum in London, but
their supplies gave out when they had no more than
begun their battle with the island jungle. They
returned to London and convinced a firm of hardware dealers that they knew the very spot where
excavation would reveal vast riches. A ship of
500 tons was chartered, provisioned for six months,
and great steel boxes with special locks were built
into the hold for the safe storage of the treasure.
Five months of frantic industry on the island found
them none the richer, except for augmented muscles
in the hardware business. Though the capitalists
finally weakened, the women remained confident,
and returned to London with the avowed intention
of finding more big hammer-and-nail men.

An excited German, who had been exploring
Cocos with five other men, returned to New York
to interest zoological societies or circus exhibitors
in a project for capturing the dragons that make
that tropical island their home. He had seen
dragons there, their footprints in the sand, their

gleaming eyes far up in the tree-tops at night, and he was sure that it would be feasible to snare a young one for zoo or sideshow. His companions, who had spent weary hours in making imitation dragon spoor and hoisting into the trees perforated cans containing lighted candles, restrained their heartless glee long enough to let him place the proposition seriously before Barnum and others.

In 1880 August Gissler was a sailor aboard a ship taking Portuguese immigrants from the Azores to Hawaii to work on the plantations. Just after they rounded the Horn, the chief engineer was disabled in a storm, and the condenser, upon which the immigrants were dependent for water, broke down. Gissler had had some experience in steam and volunteered to try his hand on the machine. After two days and nights of hard work, he got it into some sort of shape, so that it was not necessary to put in anywhere for water. One of the Portuguese offered to help him in the daily task of filling the bottles, which the thirsty steerage passengers brought to the door. The two men amused themselves with learning each other's language, and with the aid of a dictionary, were getting on famously long before they reached their destination. One day the Portuguese told a story that his grandfather had left to him in manuscript, concerning a treasure that was buried on a Pacific island, called Las Palmas. The grandfather had helped to bury it when he was one of the crew of a Portuguese freebooter called Benito. Gissler copied the manuscript, and kept it as a curiosity.

Eight years later, Gissler was living in Hawaii, having taken up some land there. A friend of his married a half-caste girl, daughter of a native woman and an old white man, whose only name seemed to be "Old Mac." His son-in-law told Gissler that the old man had in his possession a chart showing where there was a treasure, but that he had always refused to tell how he came by it. In after years, Gissler came to believe that Old Mac might have been Chapelle, but at that time he knew nothing of the involved tale of Cocos.

Gissler remembered his Portuguese manuscript. He unearthed it among his papers, the friend persuaded his father-in-law to let him take the chart, and comparing notes, they decided that Las Palmas and Cocos were the same. Sped on their way by gloomy prophecies from Old Mac, as to the evil effects on human nature of finding treasure, they sailed for San Francisco, taking with them the eleven-year-old son of Gissler's friend. From San Francisco they reached Punta Arenas, hoping to find a small schooner in which to sail to Cocos. The day they arrived, two men accosted them in English, asking if there was any chance of getting work thereabouts. They pointed out their schooner in the harbor.

"Why are you flying the Nicaraguan flag? Smuggling?" asked Gissler.

"Worse than that," they replied, "We've been looking for buried treasure on Cocos Island."

They were two young journalists from Ottawa, infected with the Cocos germ through an acquaintance who thought he knew the location of the loot. At his instigation they had bought the ship in San Francisco; when they reached the promised land, their guide took them ashore, pointed to a large hole that some one had dug, and announced triumphantly, "There,—I told you there was treasure here." Which seemed to be the sum of his knowledge. Between disgust and discouragement, they did not even look further but started back and reached Punta Arenas, penniless. They wanted to sell the schooner, which fell in perfectly with Gissler's plans, but when he went aboard, he found her nothing but a sieve.

"For God's sake, what have you been doing?" Gissler demanded.

"Pumping day and night to keep afloat," they confessed, and the last he heard of these latest victims to the lure of Cocos, they were in the interior, working in the mines.

No boat suitable for Gissler's purpose turned up. The boy came down with fever, and his father decided to take him home, leaving August Gissler to continue the expedition alone.

"If you find anything, I leave it to you to do the fair thing," was his valedictory.

At length a Swedish barque, loading cedarwood, came into Punta Arenas. She was short-handed, and Gissler shipped as mate on condition that en route to Valparaiso they should go to Cocos and stay ten days while he looked about and checked up on his information. They sighted Cocos one day, but then, like so many mariners before them, they were becalmed, the mists came down, and in the grip of the strong currents they drifted helplessly away and did not see the island again. Arrived at Valparaiso, a ship's broker was interested in the story, and eventually a company was formed and a ship chartered. The captain and every man in the crew received, instead of wages, shares in the problematical profits of the voyage. Thus Gissler first reached the spot that was to be his home, the place to which so many of his memories are bound and where his hopes still center.

In Chatham Bay the ship anchored for a fortnight. Then the captain grew uneasy; his stores were decreasing, and he saw no profits accruing here, while he knew of cargoes that he might be carrying up and down the coast. Gissler had not finished cutting his lines to survey the bearings given on his map, so the ship sailed promising to return at some future date, and left him with three sailors.

The provisions were divided man for man, those on board sharing equally with those left behind. With Gissler stayed Mike, a Dane, "who had lived with Irishmen so long he was as bad as one himself," Anderson, a mechanic, and Holm, "who was not a practical man." On the hill in the center of Chatham Bay they built a hut from some boards and pieces of corrugated iron that they found, probably the remnants of the convict settlement.

It was almost eight months before they saw the ship again. Food ran short, of course, and from boards they built a flat-bottom boat in which to row along shore to gather coconuts and try for wild pigs, which ceased to frequent Chatham Bay with the coming of men. They used to row to Nuez Island when the sea was calm enough, to get boobies and their eggs, and had many desperate times in the cranky craft among currents and sudden squalls. The hammer of their only gun was broken, and Anderson succeeded in repairing it so that sometimes after pulling the trigger a dozen times, it would go off. The difficulty was to make a pig stand still during the first eleven attempts. There were several flurries of excitement when they thought they had located the treasure, but they did less and less speculating on the subject of gold as they became absorbed in the attempt to obtain food. Also they were rapidly reaching the point where they shirked unnecessary exertion. Sometimes, tugging weakly at their roughly-fashioned oars, they would be overtaken by one of those nasty squalls which swoop down on Cocos, and then they would labor ashore at the first possible landing-place, turn the boat upside-down on two stakes, and perhaps spend the night huddled beneath it, talking and dozing, and smoking dried leaves in their pipes.

Mike was the life of the party, and Captain Gissler still chuckles reminiscently over some of his exploits, particularly in the pig-killing business, when he would stalk a sow in the tall grass, armed with nothing but a machete, which he would throw at the pig to trip it, and then, if that manœuvre succeeded, grappling with the kicking creature in Homeric combat.

One day, months after they expected relief, a schooner was sighted. She seemed to be headed for the island, but eventually drifted away and was seen no more for twenty days, when they woke one morning to find her dropping anchor in the bay. She had been sent out by the company in Valparaiso, and several of the men were those who had been there on the previous trip. The first remark of her captain was, "Where's Frank?"

He was the mate, also of the first party.

"Frank?" said Gissler, completely puzzled. "How should I know?"

When they had first sighted the island, days before, and had seen that the ship was going to have trouble in fetching it, Frank had put off in a small boat with provisions, fearing that those marooned there so long might be suffering for want of supplies. One man went with him. They had seemed to be making good progress landward for some time; then the mist had shut down, and they were seen no more. The captain supposed that Frank had reached the island days before the ship did.

He was given up for lost. Weeks later, when the expedition reached Valparaiso, Frank was the first person they saw as they stepped ashore. He had drifted for eleven days, becalmed, and bailing constantly to keep the open boat from sinking in the torrents of rain, before he was picked up offshore.

"Hello!" said Frank, "I got here first."

The next attempt on Cocos was made the following year, and the ship stayed this time, lying in Chatham Bay as before. Mike was again the bos'n. He went ashore one day to hunt pigs and surprising a large boar, fired and missed. The pig rushed at him, and as Mike excitedly clubbed his gun and swung at the animal's head, the weapon was discharged and Mike fell. It was dark before the others knew what had happened. Captain Gissler went ashore with a lantern, found him and bandaged him as best he could: a little tarpaulin shelter was rigged over him, and one man stayed to watch. When Gissler came back from the ship towards morning with a stretcher that he had worked all night in making, Mike was dead. His last words were, "To hell with the island!"

He is buried on the point of land that reaches out towards Nuez, the spot where the accident happened.

This expedition was as fruitless as the first, but Gissler saw that the location of the hoard would require a long residence, and careful consideration of every clue. He had also succumbed to the beauty of the place and was beginning to have a proprietary feeling toward this isolated emerald patch in an azure sea. There was an interval of a year or two while he laid his plans and made his arrangements with the Costa Rican government, and then he returned as a citizen of Costa Rica, Governor of the island, and owner of half of it. The terms of the bargain were that he should colonize Cocos, and that the Costa Ricans should keep communications open and give him the sole concession for treasure-hunting.   In 1894 he arrived with his wife, and six families of colonists, and they fell to work on shelters for themselves and their supplies, and to clearing ground for their plantations.   Months passed, and at last the long-expected supply ship came, bringing seven more families and no supplies! The people were landed and the ship departed, leaving Gissler with a diminishing store of food and the overlordship of thirteen families.   The land that had been cleared and planted was not yet producing much, and like a wise captain, Gissler put his crew on rations at once.   The dismayed colonists, who had expectantly emigrated to a *dolce far niente* Paradise, found themselves marooned where hard work and a food shortage made this tropical refuge painfully like their former dwelling-places.

The ship did not return; no ship of any kind came, and discontent was rife.   At last, seeing that starvation was perilously close and that they had been abandoned by the government, Gissler built an eighteen-foot boat in the little stream that empties into Wafer Bay, where the tiny settlement stood.

"With my wife's bed-sheets I made the sails," he told me, "and I went to the mainland for help," a voyage of more than three hundred miles in a very home-made craft.

The help obtained, most of the colonists departed in haste, with a revised opinion of that Swiss family called Robinson, and it was not long before Captain and Mrs. Gissler, with a peon or two, were left in sole possession of Cocos.   And so they

remained for twenty years, with three brief intervals of life on the mainland.   On one occasion
Captain Gissler heard that a son-in-law of Keating's was living in Boston, and there he journeyed
for an interview.   He found the man he sought in
hospital and negotiated the purchase of all Keating's papers, including the dictated account of his
finding of the treasure.   The information contained in them was too vague to be of use, but
from certain comparisons with other clues, Gissler
came to the conclusion that I have quoted concerning Keating's ignorance of the exact spot.

The Gisslers were not left in utter solitude during those years.   They had many visitors, all inspired with hope of sudden wealth, all with more
or less vague ideas of where to look for it, and some
much nonplussed to find a man already in possession and ready to assert his rights to the land that
was his.

In 1894 Keating's widow, now a woman of more
than middle age and married and widowed a second time, came to Cocos with a Captain Hackett
and a crew of sealers, on board the *Aurora*.   She
was looking for a spot described to her by Keating,
a spot that Gissler had already found, where a large
stone bore a carved "K" and an arrow pointing to
a hollow tree.   Hidden under the vines that covered
the trunk of that tree, he had discovered a long iron
rod, bent into a hook at one end, which was just
long enough to reach the bottom of the hollow.   But
the cavity was empty.   For days the sealers explored the island, with growing disappointment

FIG. 40.—COCOS ISLAND BOOBIES, SEEKING SHELTER AT NIGHT IN THE SHIP'S BOATS
DURING A SEVERE STORM.

FIG. 41.—WAFER BAY, SHOWING WHAT IS LEFT OF ONE OF CAPTAIN GISSLER'S HOUSES.

and resentment.   Then they accused the woman of
withholding information, and searched her and all
her belongings, without result.

Sometimes expeditions were frequent; on at
least one occasion there were two at Cocos simultaneously, regarding each other with bitterness and
exchanging accusations of unfair dealing, spying,
and destruction of landmarks.   Then months would
go by without sight of a ship, and once two years
elapsed without a visitor of any kind.   During such
intervals Captain Gissler and his wife spent busy,
happy days, contentedly cultivating the bananas,
coffee, limes, oranges, and various vegetables that
by now were flourishing, fishing the streams, and
taking long tramps over their domain, while they
discussed the millions that lay somewhere within
this narrow compass, and the possibility of finding
them.   Captain Gissler discovered the means,
which he subsequently patented, of making a very
useful and substantial brush from the natural ma-
terials at hand, and contrived a little machine to
use in the process of their manufacture.   Like all
Robinson Crusoes, he put many articles to quaint
and unexpected uses, as when he converted his
bedsprings into rat-traps that wrought havoc
among the all-too-fearless rodents.   He built a
tiny mill also, that utilized the power from the
river at Wafer Bay, and planted coconut-palms
to replace those so wantonly destroyed in centuries past.   There were wild pigs to hunt, and from
the veranda of their very comfortable house Mrs.
Gissler used a trout-rod every day at high water,

casting into the broad pool that rose over the sandy beach.

Smoothly the days wheeled by in a procession of soft winds, hurrying rain-clouds, moonless or silver nights, flowing into unregarded weeks and months of perpetual summer. The crash and suck of breakers, the rattling song of the wind in the palm-trees, familiarly became a part of silence, and time was counted, not by dates, but by events,—the day the boat broke loose, the last big rain, or the evening that a light glimmered on the horizon. When treasure-hunters came, there was the news of the world to hear, and gifts to exchange, flour and sugar for fresh pork and coconuts, and always the game of twenty questions that the newcomers played in their attempts to extract useful points from the oldest resident of Treasure Island.

Not all the visitors conducted themselves with courtesy. In 1896, while Gissler was absent in Costa Rica, renewing his arrangement with the government, a British warship dropped anchor at Cocos. She was commanded by a somewhat impetuous Irishman, and I will follow the cautious example of another writer in calling him Captain Shrapnel. He had a romantic soul but bad manners; he landed three hundred blue-jackets, informed Mrs. Gissler (who was alone except for two peons) that she was not to leave the immediate vicinity of her house, and disregarding her protests that this side of the island was her husband's property, turned his men loose in a three-day orgy of blasting and drilling in search of the loot of the

Mary Dear. For this exploit he received a reprimand from the realistic Admiralty, and Cocos was put out of bounds so far as the British Navy was concerned. The First Lord perhaps visualized the island as exerting an influence like Sindbad's Magnetic Rock, and drawing in Her Majesty's ships one by one.

Captain Shrapnel had left Cocos before Captain Gissler came back from the mainland, but the idea of buried treasure had a firm grip on his imagination. Some years later he returned, having arranged the financing of a civilian expedition, and spent some busy weeks round Chatham Bay. He had received information from a man named Fitzgerald, to whom Keating had confided certain directions on his death-bed. These bore such fanciful touches as the face of a cliff that would, when a spring was pressed, revolve and reveal the treasure. Whether it was Benito's crew or that of the *Mary Dear* that included such skilled stone-cutters, does not appear to have been explained.

There was also Lord Fitzwilliam, a wealthy British peer, who tried his hand at treasure-seeking. He stole a march on Captain Gissler by beginning blasting operations at Chatham Bay without the owner's permission, but a large section of cliff fell on his head and possibly persuaded him that the God of Buried Treasure fought on the side of the mannerly.

Another English expedition was financed by a Mr. and Mrs. Gray, who came on their yacht to this island whose soil has been so industriously

tilled. Gissler battled more than once with the
Costa Rican authorities, who granted concessions
to other treasure-hunters, forgetful of, or disregarding, his prior claims.   People are always
springing up with something that they assert is
infallible information about the Cocos hoard.   A
stray newspaper item will call forth letters boast-
ing of secret clues, obtained in mysterious ways.
A few years ago the casual announcement that a
professor from a mid-western university was going to the South Seas in search of museum material
brought him a letter from a man in Maine, offering to obtain for him a chart of Cocos showing
where he could find sixty million dollars.   Those
who speak of Cocos never stint themselves on millions.   There was even an attempt to find the treasure with a divining-rod, and this year the papers
have told of the latest search which is to be conducted by an Englishman, "using the latest scientific inventions for finding buried treasure."   If the
reporters quote him accurately, (which is open to
question) it seems doubtful if he will even find the
island, as Cocos seems to be confused with Cocos-
Keeling; he might look in the wrong ocean, since
the location of the treasure island is given as "on the
fringes of the sinister Sargasso Sea."   At any rate,
the modern scientific developments will be interesting.

It was not until the *Arcturus* returned to New
York from her six months' cruise that we discovered Captain Gissler. We had heard of him, of
course, as has everyone who knows anything of the

tale of Cocos, but he seemed the sort of legendary figure whom one could never hope actually to see. We had visited his settlement at Wafer Bay, and seen his house, whose crumbling piles have let the structure settle crazily to the ground, the remains of storehouses, the little iron stove sitting forlorn and rusty at the edge of the beach where it has been dragged by some visitor since he left. From his plantation, all but undiscoverable in the tangle of wild things that are springing to reclaim their ground, we picked quantities of limes, and carried them aboard in an empty box marked "DYNA-MITE."

But it was months later that I climbed those creaking New York stairs, and knocked upon a door. It might have been the cover of a Conrad novel, for it opened on a figure that he would have understood and interpreted as no other could.

A big man, straight and upstanding as a youth, with a white beard that covered his chest, bright blue eyes that could twinkle or glower, and the shipshape trimness that speaks of seafaring, opened to me. Four words—"I've come from Cocos"—were my magic formula, and in less than that many minutes we were in the midst of the story. Over an immaculate cloth we discussed delicious coffee, while the tale of treasure unfolded and the deep-toned mutter of the town seemed to change to the rush of the precipitate storm down the deep ravines of that island that has not yet given up its secret.

CHAPTER XI

THE PHILOSOPHY OF *Xesurus*

I HAD made probably forty submersions in my diving helmet, and on my last ascent sat shiveringly on the dripping thwart and with water-wrinkled fingers scrawled damp notes on what I had seen. About this time I became obsessed with an unendurable impatience when I thought how relatively little of cohesive value I had obtained during my two score descents; what slight correlation I had observed among all the submarine activities. I tried to parallel that day's notes with corresponding items which a Martian, dropped into Fifth Avenue or Regent Street, might glean in a few minutes' time:

Descended eighteen feet and sat on a volcanic block as large as an automobile, covered with great round patches of orange and purple sponge; little fish swam curiously around me and dived into grottos just out of reach. I could not move about much for there were patches of long-spined sea-urchins everywhere. A school of small ladyfish and wrasse came to a bit of crab-meat which I

Landed on the edge of a machine as large as a small ether-cycle, with glaring posters of strange, beautiful women and the place of the murder, plastered on its side; newsboys crowded around and swarms of people dashed into holes in the ground, and poured out of places called Exits. There were spikes on the top of a park wall and Keep Off signs so I could not walk about freely. Some held. Twice a great hiero- sparrows hopped up and ate glyphic fish poked his head crumbs at my feet. A curi- out of his crevice and rolled ous old man opened a window his eyes up at me and then across the street and peered a golden grouper swam slowly down at me once or twice. by, like a wandering ray of Two lovely ladies passed but the sun. A fish new to me was did not look at me. One of a large wrasse, green in color, them had on a most wonderful with two longitudinal black sea-green dress shot with glints lines, broken up into elong- of wine color, which came and ated dashes. A tiger shark went in the murky sunlight. watched me suspiciously, and A little distance away a po- came so near that I stood liceman watched me intently, up and took hold of the lad- and then came toward me with der, although I knew I had such evident suspicion, that I really nothing of which to be rose slowly, yawned, stretched, afraid. and walked slowly away.

Observations such as these, while having an accumulative value when sufficiently numerous, give little or no idea of a complete picture, or well-rounded appreciation of any group or individual. My Martian might better have concentrated on some artisan or laborer, or any interesting person whose dress and actions and general life revealed some fundamental purpose, or method, or reason of existence—of reasonable relationship to all the host of objective phenomena which composed his environment.

I made up my mind that the next time I dived, I would bring back the image of a personality, the *raison d'être* of some fish. That afternoon the first fish which caught my eye when I reached the bottom rung of the ladder was a yellow-tailed surgeonfish, and I seized upon him to point my moral and adorn my tale. It was a literal seizing, for I harpooned him forthwith and carried him

up to the well in the boat. He was one of about
seven or eight hundred which were so busy grazing that they paid no attention to the abstraction
of their comrade. These blue cows, as we called
them, are fish from a foot to eighteen inches, weighing from one to four pounds, and are by far the
most abundant of this medium-sized class. Their
body is very deep and compressed, like most of
the surgeonfishes, and their thick, pouting lips
and protuberant eyes make them look absurdly like
some stout people I see from time to time.

Ninety years ago the French frigate *Vénus*
paid a visit to the Galápagos Islands and a specimen of these yellow-tails was collected. To this
specimen Valenciennes gave the name *Prionurus
laticlavius,* but the exigencies of priority demanded
a shift and today it is known as *Xesurus laticlavius.* It is an appropriate title and freely translated means the Side-striped Scraping Tail.
Using Tail as a proper name is excellent in this
case, for of everything about the fish the tail is
the part most conspicuous (Plate VI).

They are a uniform slaty-blue in color with two
broad bands of black which extend downward
across the body, one beginning on the neck and
curving downward through the eye to the mouth,
and the other just behind, starting at the front
of the dorsal fin and ending at or below the pectoral. The tail is bright greenish yellow, and a
streak of this color reaches forward beyond the
base, outlining, on the sides, the three poisonous,
file-like spines.

The great numbers of these fish show that they are successful wagers of life, and their conspicuous pattern and coloring combined with their absolute fearlessness indicate that they have some adequate defense against the creatures on every side, who would gladly devour them. The mouth is absurdly small, with wholly inadequate teeth, as far as biting is concerned, so as to that method of defense these submarine cows are on a par with the grazers of the land. They have no long, strong tail to lash, nor have they the static defense of the funny little box-fish, and their flesh is not at all poisonous, but delicious eating, as we proved more than once. So we must fall back upon the caudal armature as the crux of the matter.

The surgeon or doctor fish show a beautifully graded series, from a form which has a long, curved lancet, sharp as a surgeon's scalpel, folded forward into a groove on the side of the tail, to others, at the opposite end of the scale, which have only a shagreen-like roughness of the skin. *Xesurus* lies not far from the lower end of this series, meaning, by lower, the more primitive condition. It seems probable that this whole group is descended from some form which had, as a defense, the entire body covered with bony plates, from the center of each of which arose a curved spine. In *Xesurus* these have degenerated until there is left only an irregular group of ten to fifty small, black, dermal plates, scattered over the posterior third of the body. The most anterior are mere spots of dark pigment, then a minute, central, rounded nodule appears, shiny and black; this increases as we go farther back, and a good-sized basal plate develops; the raised cutting edge becomes sharp and horny white and an anterior hook appears. At this point, however, there is an abrupt transition to the three, large, caudal plates, one in front of the other and separated by about their own width, each of which supports a hooked file.

A complication, although not a negation of this theory, is that in the young *Xesurus,* less than an inch in length, the spiny ridges are said to be very low and serrate, and the irregular scattering of plates on the body is not discernible, developing only later in the life of the individual. I have not seen a yellow-tail of this age, so am unable to confirm or deny this statement, or to tell what careful dissection of the skin might disclose.

Much has been written about these "murderous, poisonous" spines, but as far as I know, no definite experiments have been made as to the latter quality. In the first place the defense of our *Xesurus* is comparable rather to irregular, sharp-ridged, hooked files, than to spines, so that there is no possible chance of actually disabling any assailant large enough to kill and eat them. Even the force of numbers can be of little avail in any initial attack, and eight hundred *Xesurus* crowding about an attacking shark or barracuda could do little more direct harm than hamper his movements and partly smother him.

I made four experiments to prove the venomous quality of the mucus about the spine or any liquid which might be operative in connection with it, and I obtained decidedly positive results. Other more elaborate experiments had to be abandoned during this expedition. I took a live *Xesurus* and armed with thick gloves I bent its tail slightly around and rasped the sharp files against the scales of three species of fish, one a much larger form, *Seriola dorsalis,* and two smaller than the surgeon-fish, a *Pomacentrus arcifrons,* and an *Evoplites viridis,* both of which live in the same locality as the *Xesurus.* I had no large carnivorous fish, but it is unlikely that the results would have been different. In each case I had a number of other individuals of the same species, as controls, all living well in our aquariums. I watched the fish carefully but after the excitement, due to my taking them from the tank, was over, I saw no symptoms of discomfort, the abrasions themselves being quite negligible. The following morning all of the four subjects of the experiment were dead, their fellows, without exception, being still in perfect health. There was a slight discoloration of the flesh about the rasped wound, but no other lesions.

In the case of a butterfly protected by nauseous juices, every inexperienced bird and lizard has probably to catch and taste for himself—the race of butterflies winning immunity at the sacrifice of one of their number. Turning to the life and death problem of *Xesurus,* from a general point of view there seem to be only three methods of correlating the various possibilities and factors. Corresponding with the case of the butterfly and the

lizard, we must (1) imagine every shark and barracuda, moray and grouper as taking toll for himself, and furthermore that the action of the spines
and poison is, in their case, only an exceedingly
disagreeable and distasteful, not a fatal one; or
(2) we must believe that every assailant is poisoned and dies immediately, when the result would
be simply, how soon all the sharks and groupers
would be dead from eating surgeonfish; and (3)
we may imagine an instinctive knowledge of the
dangerous qualities of the yellow-tails on the part
of sharks and others, induced by the gradual elimi-
nation of xesuruphagus individuals.

Once the tremendous interest of this problem
became apparent, I was always on the lookout for
some hint of a bout between these grazing cows
and their enemies, but never did I see a menace
or a defense. Their lives were lived calmly, with
dignity, and wholly superior to the terrors and fears
which marked the movements, the activities and the
habits of most of the fish around them.   Their
cousins, the surgeonfish with long, sharp, wicked-
looking spines were never as abundant or fearless
as these, although one would say they had a much
more effective means of defense.

Another problem, quite as difficult of elucida-
tion, has to do with the near relations of the surgeons, the Chaetodonts or butterfly-fish, and
Balistids or triggerfish. So intermixed are the
characters of these three groups,—characters ex-
ternal and internal, both of the body organs and
of the skeleton, that systematists group the sur-

geonfish sometimes with one, sometimes with the other. There seems, however, little doubt that the butterfly-fish and the surgeons are much closer. A matter of some fifty million years ago, in the Eocene, there swam a family of fishes—the Pygaeidae—in which these two groups were brought very close together indeed (Fig. 43).

We need here concern ourselves only with the character of the mode of defense, which is curiously different in the three living groups. In general, that of *Xesurus* sets it rather apart from the others, whose dependence is upon the anterior spines of the dorsal fin.

In the butterfly-fish these are very long and strong, but not especially modified, and grade into the posterior, lesser spines of the dorsal. Both in the aquarium, where we kept black-fronted butterfly-fish alive for several weeks, and near the bottom of the shallow shores where these exquisite fish lived their lives in pairs, I watched them fence. The simile which comes to mind is of a pair of full-grown tahr on a steep Himalayan mountain side. I have watched these splendid wild goats through long-range glasses, each shifting into most graceful poses as he feinted and made passes with his horns, either in play or in grim earnest, kneeling, swinging sideways, rearing lightly into the air with forelegs bent under and horns playing like rapiers.

This morning when I was diving, a large sea bass passed close to where a pair of butterfly-fish was feeding, and as it approached, one of the two went out to meet it. Every great spine gradually

rose into place, like the nuchal fringe of any amblyrhynchus, and as the annoying fish did not swerve
aside, the dainty yellow and black *Chaetodon* went
through a hundred graceful threatenings, rearing,
ducking, dipping far to one side, and making swift
passes at his opponent, bringing the whole body
into play.   Now and then it would jerk upward
with all its force, with an unexpectedness which
the other fish could only just manage to avoid,—
and which, if it struck home, would work real
damage.

In the triggerfish, the anterior dorsal spines
have become quite as specialized a means of defense
as the spine of the surgeon.   Complete detachment
has been brought about from the functional part
of the back fin, the first spine being long, often
serrated, and usually held in place by a second
smaller one (Fig. 46).

Leaving the family relations and the devious
and obscure ways by which the yellow-tails have
won and are holding their present enviable position, let us consider the details of their fitness for
the everyday labor of life.   Our name of cows
was given because of their everlasting grazing,
nibbling, nibbling, nibbling, at the plant and
animal fodder which covers the rocks.   The habit
of going in such enormous schools, and crowding
closely together made them a spectacular feature
of every island where I dived, and their manœuvers were astounding.   Several hundred approached
swimming slowly along, when, as if at a signal,
all would stop, and if over a rather flat bottom,

YELLOW-TAILED SURGEONFISH
Xesurus laticlavius (VALENCIENNES)

(One-half natural size)

PLATE VI

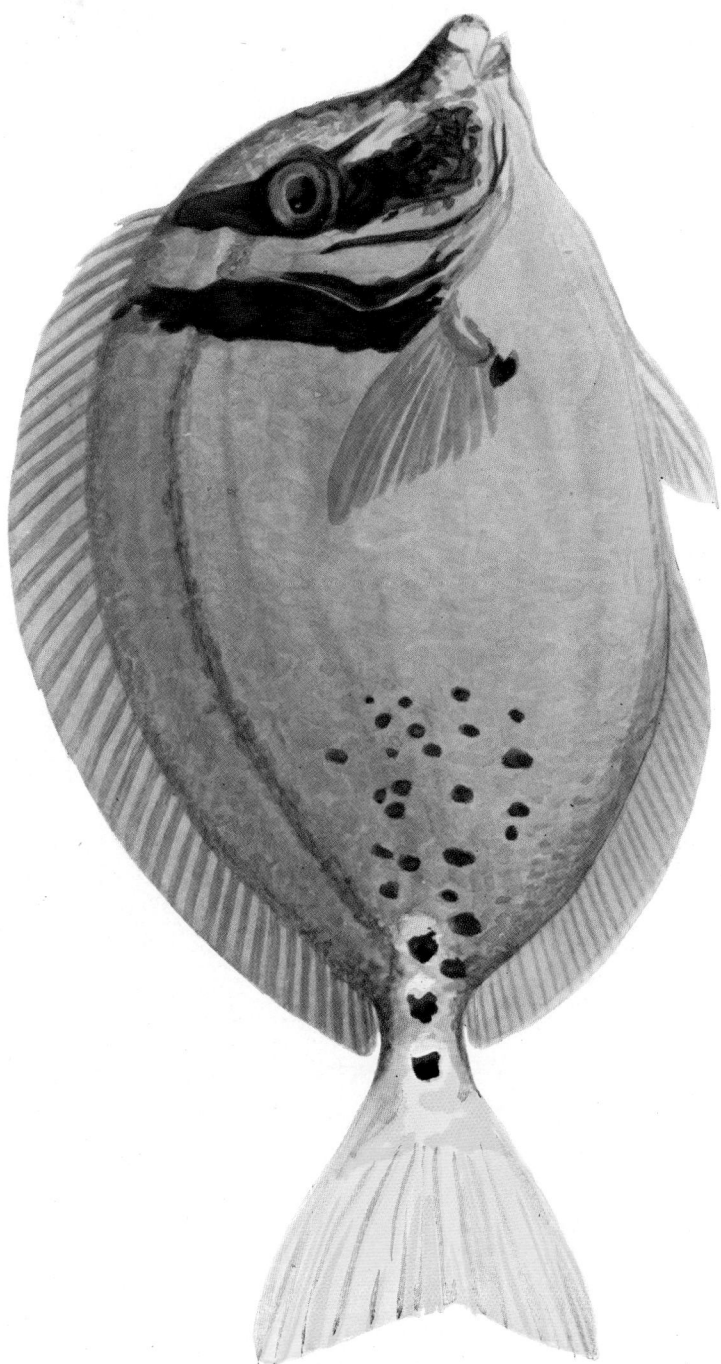

would up-end like ducks, and begin to graze. From a long, crowded mass of blue fish, they changed, as one, to an army of banners—a maze of fluttering, golden flags, all crowded close, all furling and unfurling, lighting up the flat spot where the surgeons fed, as a clump of goldenrod will catch and glorify a sun's beam, and toss it back to rejoice our eyes.

As I have said, they were the most fearless of all the fish of these waters, and when a few moved over to look at me one by one, all the rest shifted, and the first had to move on, if only to make room for the scores pressing up. Once when I was surrounded by a herd of yellow-tails I chose a comfortable seat, and deliberately studied their architecture with appraising eyes. Every line and profile and character seemed a perfect adaptation to their feeding habits. The high, compressed body, almost surrounded by fins, with an extremely mobile, caudal peduncle, allowing the tail to turn at right angles to the body, all helped to sustain, or to shift the fish quickly against the surge or to hold it steady while the grazing went on. I never realized so fully the stiff, immobile quality of the whole body of the fish. It could roll its eyes, twist its tail and bend very slightly, but the teeth and jaws were without other than vertical movement. The entire lack of a neck made it necessary for every fin to help with each bite, pressing and holding it firmly while the teeth scraped and closed, then drawing back slightly, while the food was ingested and swallowed, immediately shifting slightly to one side or below, and ahead again for another scrape. I was able to analyze these successive movements, but in reality they followed one another with the swiftness and ease, the precision and correlation, of a man's steps.

The mouth was strongly protuberant, the jaws being wholly beyond the normal curve of the forehead, nostril and chin. The lips were soft and attached so far back that they could be drawn out of the way of any other part of the face or mouth. The teeth were perfectly adapted to their work— remarkable little scraping machines which cleaned the growths from the rocks as a hoe cuts the weeds from sod. They were the strangest-looking teeth in the world and at first glance recalled a double row of the tiny ivory hands on long sticks which the Japanese carve so exquisitely. Under the careful scrutiny of a lens, another absurd, and this time a perfect, simile forced itself upon me. There were nine on each side, both above and below, thirty-six in all, and to the smallest curve they were not like hands, but feet—thirty-six little soles, with five, well-graduated toes on the tip of each, a graceful in-curving arch, and a delicate heel. The teeth were inserted at a strong outward angle, and overlapped on each side so that the functioning top of each tooth was limited to the great, and to the next two toes. For a time it was difficult to be abstractly dentistic in my contemplation, and not laugh at the thought which the eye compelled, of eighteen little men just disappearing down the throat of every *Xesurus*. The rounded tips were evidently ideal scraping organs, and my comparison with a hoe had better be replaced with that of a rake.

The nostrils were far out of the way near the eyes, for they can be of slight use, there being little or no selection of the scraped-off nourishment, and if placed nearer, they would only become clogged with debris in suspension. The eyes were of good size and very protuberant, standing well out above the surrounding, rather concave head area. Their rotating power was unusually great. With a normal divergence of 8° forward, and 12° down, which alone focussed them well toward the approaching rocks, they could be rotated forward and downward through an angle of 42°. Not only was the elevation and direction of the eye thus a specialization for the direct observation of the feeding grounds, but the cheeks were hollow, and the elongated bridge of the face deeply concave, thus affording an unobstructed field of vision. This was another powerful argument for the absence or relative lack of enemies, that all this complicated architecture was for clear vision ahead, not behind, —*Xesurus* was in no sense a pursued one.

The pattern and coloration were not protective, and if they were, the enormous schools would render any concealing coloration of no avail. The pale blue grey, with the two broad, black bands, the large, silvery iris and whitish lips, and above all the brilliant yellow tail, and yellow and black-banded line of the caudal plates, rendered it an object easy of detection among the variously colored rocks on which it fed. I should rather

classify the tail as a recognition mark, a feature
so characteristic of gregarious animals, and the
lateral parti-colored band may, as likely as not,
have served the purpose of a warning signal to
any who may have instinctive appreciation of the
danger which it advertised.

The gill openings were enormously elongated,
perhaps as an aid in permitting a strong in-and-
out-rush of water, when this was roiled by floating
detritus from the continual gnawing. Just back
of the gills were the pectoral fins, long, with a hint
of falcateness, and properly strong to govern the
myriad adjustments of every day's activity. When
balanced close alongside a rock, the pectorals were
used alternately to fend off with, in addition to
their more usual functions of balancing and pro-
pulsion. They are wonderfully strutted, hinged
on an oblique base, which constitutes the cross-bar
of the A formed by the clavicle and post-clavicle.
The superficial muscle which leads straight forward from this fin, controls the posterior half of
the pectoral rays, and, when it contracts, curves
them around and out, until they form a most symmetrical, forwardly directed trough or cup—which
in its function of a brake or backing organ, is of
more importance than the backward push of a
swifter fish.

Below the pectoral and slightly to the rear,
the ventral fins arose, close together, on the profile
of the chest. They were fronted by very stout,
rough spines, and the chest directly in front was
quite broad and flat. A number of times I saw

individuals propping themselves for a moment on this part of the body, the ventral spines acting as two legs of a tripod, or again pushing hard against the rock when the fish slid over a sharp angle.

Whether from disuse or lack of incentive, these fish seldom exhibit any burst of speed, although the dorsal and anal fins were long and deep and the tail and the pectorals large and powerful. When moving along or feeding, I never saw an unusually swift movement, and even when I was harpooning them, and now and then thoroughly alarming them, they never showed more than an ability to avoid an awkward thrust of the grains. I could easily spear a half dozen of them to any one of another species, and not because of their abundance.

Another argument in favor of the lack of enemies was the very considerable variation existing among *Xesurus*. In a large school I saw some which were exceedingly deep in the body, and others a full third lower; the lateral line might be present or absent, and comparison of pectoral fins of different fish showed very marked variations in size and outline. Fish which live very strenuous lives, whose numbers are kept down to low limits and which are beset by numerous enemies, exhibit little variation from the normal,— they keep to the narrow, sharp line of sheer existence and every character tells—any latitude in one or another direction might well wipe out the whole race.

Again and again they came to the crab meat, but I never saw them nibble at it. The attraction seemed only that of curiosity—they were like city strangers looking through the window of an automat. They fed at all hours, and twice at night by the aid of a water-glass and my electric flash, in shallow water, I have seen a small school scraping away as though it were day or at least moonlight. In this noncurtailment of meal hours they differed widely from carnivorous fish and resembled their dietetic relatives—sheep and cows.

Once, and once only, I took one on a small hook, baited with crab meat, so I suspected they were not wholly vegetarian in their scrapings. I sought confirmation in the examination of a number of stomachs. About sixty per cent contained solid masses of green, succulent algæ, and in the others there were in addition bits of rock and shell, and remains of crabs, shrimps, sea-urchins, worms, and all the odds and ends of animal life which find shelter in the short seaweed fur of the rock surfaces.

The viability of the yellow-tails is very high, and I can recall no instance of a harpooned fish failing to live and thrive in our aquariums when restored to running water immediately after capture. None, however, long survived the tainted waters of Panama and Colon, combined with the enforced lack of running salt water during the passage through the canal.

CHAPTER XII

SLUMBERERS OF THE SURGE

WHEN I began to be wonted to the long, winding kingdom of my shallow, underwater world, its strange landscapes and stranger inhabitants slowly penetrated the first fine frenzy of inarticulate emotion, to more specific appreciation. And the very first evidence of this was humorous—for I began to see close resemblances between the villagers of the deep, and dear friends of mine. And this is not to be read with a roar of laughter, and an all-inclusive pseudo-witticism of queer-looking people and "poor fish." That is far from what I mean; it was in no way a question of special features or personal appearance, but often in quite indefinable qualities. The way a grouper would come over a mound of coral, or a moorish idol peer up at me, the nervous flick of a small wrasse person, brought often to mind a gait, a glance or a trick of the hand of someone. These casual chuckles undermined the distraction of alienness, and at once I felt more at home. This was emphasized when I dived again and again in one spot, day after day, and saw, not only the same

lanes and streets and mountains, but the identical
fish themselves.   The little old lady in Paris,
garbed in black, who used to pass me on her way
to market every day had always the same tear in
her veil, and now, the small, fussy demoiselle fish-
let which invariably scurried past when I had
taken my seat, was known to me among all her
neighbors by the frayed spot on the side of her
fin.

I succeeded in merging myself with the life of
fishes, aided by the lack of fear or even respect
with which they greeted my entrance into their
world.   But when I began to think in words I
found that just as I had to have a stream of atmosphere flowing down to me, bringing with it all
the little motes and beams belonging wholly to the
upper world, so when my mind began resolving
what my senses sent to it, into outflowing words,
these were ever burdened with dry-earthly similes
and metaphors.

To an eye above the water my new kingdom's
limits, within the confines of these Cocos and Galápagos Islands, would appear like a multitude of
thinnest of rings scattered about just beneath
the surface.   For this is an egocentric kingdom as
far as I am concerned, and its lower boundaries
are those of my pitiful extremes of penetration.
As for the upper frontiers, I admit neither rock
nor weed ever bathed by the air even at lowest
tide.   All between I have made mine by right of
imagination and a few score of timid entrances
and creepings about.   Yet always, while among

my subjects, I must abide in a glass house, and like a humble water beetle enclose within it a bubble of air. My impatience never relaxed the desire to fling the glass windows wide open, and smell and taste and hear this new world—to *hear,* for there must be some rippling vibration of sound or other waves from so many thousands who forever mumble at one another with their lips.

One of my favorite neighborhoods of observation was a marvellous shire on the bottom of the east side of Chatham Bay, Cocos. Just as Cocos itself at this season was more often than not completely cloaked in a solid rain cloud, so my capitol was forever hidden from prying eyes by a liquid sheet of emerald green.

Before describing an earthly city, we always speak of its environment and background. What I saw as I looked around above water just before I dived was a sort of upground, I know of no other word—the beautiful, great bay with the *Arcturus* riding at anchor, while high overhead rose the steep mountain slopes of Cocos, covered with dense, green jungles—tall palms and graceful, lace-like tree-ferns standing out above all the rest, while fig trees clung to the steepest slopes, dropping down perfect portières of dangling rootlets. In and out, like a warp of silver threads among the green foliage, shone the waterfalls—the glory of all this island loveliness, dozens of them, slipping down from rock to rock, or sliding gently over hundred-foot stretches of emerald moss.

But now the helmet is poised on high, dropped

over my head,—I am eclipsed, and change planets.

I sink down, down, down, and finally let go the last rung, drop quietly and deliberately on my feet, and look around at a city of giant mushrooms. A huge dome in front of me offered good climbing, so I kicked my feet and body free, and drifted to the top with slight tugs of my hands, gravitation all but negatived. From the top I looked down upon a marvellous boulevard of the whitest sand, bordered by edifices of coral beyond all adequate adjective and exclamation. In the middle distance I saw the palace of the Dalai Lama at Llhasa with its majestic down-dropping lines, beyond it the corals had wrought a fairy replica of the temple of the Tirthankers at Benares. Then a cloud of pagodas filled the end of the sandy vista, silhouetted against the blue at which I can never cease marvelling whenever I think of this water world,—a pale cerulean, oxidized now and then with the glimmering through of some still more distant monument. Invariably the architecture of the East was brought to mind, not the semi-plagiarized structures of most of our western efforts, but light, uplifted pagoda roofs, curving domes, and stalagmite minarets, together with the scroll-work which is lace-like but never gingerbready.

Through many days of watching, sometimes rising to the surface grey with the soaking of water, or chilled and chattering, but always reluctantly—I studied the fishes, the aborigines of these places, and I found them astonishingly like humans in all their more important habits and concerns of life.

Looking over my finny subjects in general I found they were divided into distinct gens or castes, and these in turn separated more or less naturally into guilds and professions. From my seat at one end of the mushroom city I could pick them out—sometimes several at a glance. Over the coral, above its mounds and branches and labyrinths, there floated the castes of Free Nomads and Grazers. Shall I call them figuratively the zeppelins and the airplanes of the sea, or, with rather more exactness of applicability, the eagles and vultures, the parrots and woodpeckers? Or, best of all, let us credit them exactly for what they are.

As Nomads I should consider those fish people who usually hunt singly, but sometimes in small packs, who have no homes, no coral haunts or rocky retreats, but who live, feed, fight, mate, sleep and die in mid-water. The sharks are these, but not the rays and skates, which belong to the same natural order, but which have spread into various directions and appropriated an interesting and profitable field for themselves. Indeed, in the case of the sharks, what has not been usurped by them has been given them as endowment by legend and fancy. We humans adore to build up a scarecrow of straw and paper around things admirable in themselves, inflate it with hot air, then look at it, scream, and run terrified away. Cries of *Snake! Evolution! Shark!* are sufficient to throw certain

panicky, timid souls into a horrified terror. All of
these fears have about the same basis of truth; out
of seventeen hundred species of serpents living
on the earth today, less than one third are dangerous; undoubtedly there have been a few men
who, at the same time, have been very bad men and
believers in evolution, and there is no doubt that,
since history has been recorded, a few authentic
cases of the attacks of sharks upon men have occurred. To condemn sharks in general is like
never taking a taxicab because men have been run
over and killed by taxicabs.

I have written elsewhere of individual sharks
I have met, but here we are concerned only with
their relations to the scheme of the shallow water
world. At Cocos, there weaved in and out above
me, occasionally coming down and curving around
the great coral pagodas, sharks of three species.
The white-finned and the island sharks were wandering nomads of clearly vulturine habits, arousing
no fear among smaller or weaker fish, but always
on the lookout for a crippled or dead creature.
They were the dominant scavengers, and after we
had used dynamite, the sharks under water and the
frigatebirds above, cleared away every overlooked
specimen, no matter how small.

These two kinds of grey sharks were four to
nine feet in length, and they swam slowly, with
wide lateral undulations of the head and body,
keeping rather a dull outlook from their yellow
eyes. The ability of the human imagination to see
what it thinks it ought to see is astonishing. As long

as my book-and-legend-induced fear of sharks dominated, I saw them as sinuous, crafty, sinister, cruel-mouthed, sneering. When I came at last to know them for harmless scavengers, all these characteristics slipped away, and I saw them as they really are,—indolent, awkward, chinless cowards. They are to a barracuda as a vulture to an eagle; a ladyfish has a thousand times less weight and double their courage.

As regards tiger sharks, which, by the way, attain a length of thirty feet in my kingdom, I reserve judgment. I have had medium-sized ones swim up to within six feet and show signs of nothing more alarming than curiosity, but I have also seen a tiger shark snap up a baby sea-lion close to a rookery of big males, as though it were a minnow, and I have observed and shared the respect with which fish sometimes greet his appearance. I should catalogue him as an uncertain character— safe enough usually, but to be interviewed with the iron ladder between us.

Groupers are another tribe of Nomads, one without any sense of humor, or the sophisticated casualness which seems to me to characterize most sharks. Groupers take life in grim earnest and while they lack the pessimistic viciousness of barracudas and morays, yet they are persons of uncertain temper. Lack of size alone keeps them from being as much feared as tiger sharks. I was never wholly comfortable when these great brutes came up in their loose schools of six or eight, swimming so close that I often kicked at them or stabbed with my harpoon. Their reaction, after avoiding the stroke, was instantly to return and follow the foot or the instrument in a most disconcerting way. No shark was quicker, nor by a long way as effective in attack upon any fish in trouble or disabled, as these evil-mouthed fish.

Once I saw a giant ray or devilfish while I was perched on a coral throne. Dense schools of small fish passing overhead dimmed the light as would a cloud, but this huge creature actually caused a momentary eclipse as he flew close above me, so close indeed and so far beneath the top of the water that my companions did not notice him. My delight at seeing his enormous enamel-white expanse overhead was temporarily distracted by one of his wing tips catching in my hose of life, but it slipped around with no more than a sudden twitch to the helmet. I entered him in my census file as Nomad, unique so near shore, harmless, curious, playful, feeding on nothing more exciting than the minute shrimps and infant crabs which paddle about near the surface, especially at night. All this I had gleaned from others of the kind which I had met farther out in the bay and elsewhere and always close to the top of the water. Devilfish he may be in appearance because of his horns and tail and the color of his cloak, but he has a gentle soul. This giant must have a courtship of sorts and a consorting for a time with a mate, but I have found him the most solitary of behemoths.

The last member of the tribe of True Nomads, and far and away my favorite, is a splendid blue carangid, two feet or more in length and swifter than any other. In clans of ten or a dozen they come out of the translucent blue, and, as they approach, slip off the azure veil which dims them and flash out pure silver, for, from my position, I see them with the eye of a true dweller in these deeps. They are built with the finest of stream lines, narrow peduncle aft, wide crescentic tail, long, falcate pectorals. Around and around me they go, arousing keen interest and admiration, where the groupers induced suspicion and distrust. I felt that these were fish of caste, fighting, if they must, in the open. Their relation with other smaller fish was a mystery or else to be explained by sleight-of-fin legerdemain. None paid any more attention to them than to the grazers. Yet these were of a far other sort. Three separate times I saw one of these carangids move out of the circle they were drawing around me, with a twist and a flash as quick as light, and each time a small wrasse swimming near, absolutely disappeared as if suddenly dissolved. It reminded me of the frog-and-his-tongue trick,—a frog facing a fly a considerable distance away, and suddenly the fly is gone. You are sure it went down the frog's throat, but no human eye is quick enough to see all the details. And so the flash of silver caranx seems not to approach or touch the little wrasse—and yet the wrasse is no longer interested in food or life, and the caranx is back in place, swimming quietly, breathing gently.

These fish would take no bait and they avoided

the repeated stabs of my grains with less than effort, and only when we took advantage of them with a stick of dynamite was I able to name them for certain, *Caranx melampygus,* and to study their marvellous body engine at leisure. I learned as much as any instantaneous cross-section could provide, of which one fact only is of interest here. A female with ripe ovaries was about to deposit one hundred and fourteen thousand eggs. This showed clearly that no matter how well able the full-grown fish were to take care of themselves, yet the young fry must be threatened with a host of dangers to render such a number of eggs necessary to maintain the species. To return in this connection, for a moment, to another Nomad, it is thoughtful to consider the devilfish which produces but a single young, weighing nearly thirty pounds at birth. On this very trip I examined such a lusty infant and could see no means of defense by which it could escape the attack of a barracuda or tiger shark. I should like sometime to take a year off and do nothing but study the life history of the devilfish.

Once when a boy I was studying the common bird life of a small city park. I looked up one day and saw a brilliant parrot perched in a tree overhead. The thrill which came to me then was repeated when, almost on the last day of my diving at Cocos I saw a beautiful flyingfish swimming over my mushroom coral city. I had hardly registered it when the reason for its presence in this unlikely spot was explained. A long, narrow fish

came up behind, slowly at first, then with a rush—
a needle-toothed garfish. The flyingfish gave two
or three convulsive surges forward and then I saw
what I had never expected to—one of these fish rise
from the water above me and disappear into the
air. Somehow this made me feel more like one of
the actual inhabitants of this underworld than
anything which had occurred heretofore—I was
seeing things from a real fish-eye-view.

The gar missed his prey and I was interested to
see that he became utterly confused, and made one
short rush after the other in various directions. I
saw the flyingfish drop into the water only twenty
feet away, coming into view with a flop. The
gar showed no signs of having sensed this, and the
last I saw of the two, the pursued was vanishing
into the blue distance while the gar turned back
the way it had come.

The last Nomad I can recall really does not
belong in this class, since it has a home, although
the strangest in the world. When the devilfish
swam over, I saw very distinctly, two sucking fish
glued to its under side. These are the remarkable
attendants which spend their whole life being carried about by their host, whether shark, devilfish or
turtle, so if not comparable to the nomadic Arab,
they can at least qualify as the representative of
the Arab's flea.

I stood up on the top of the great coral mass,
which I might, if I were that kind of person, have
named "Nomad Belle Vue," and slipped, or
rather drifted half-way to the bottom. Then with

a mighty spring I passed slowly but quite across
the sandy boulevard and beyond to another city,
this one of cones, inverted cones at that, like enormous anemones. The swell was increasing and far
off at one side I could see the iron ladder frantically
jerking up and down. Hardly had I curled myself in between three small cones, with a branching
tangle of animal blossoms in front, when there
occurred that dimming of the light with which I
had become so familiar. Leisurely there passed
overhead three hundred—three hundred and
twenty-six to be exact—of the big, black surgeonfish. They wandered over to the great brown
coral which I had left, and spread over it like a
herd of sheep across a meadow. Nibble, nibble,
nibble, as they climbed slowly, drawing the black
blanket of their numbers over every inch of the
surface. A strong surge swept them a yard away,
held them suspended for a moment, and then returned them each to his place on the coral.

To my coarse and untutored vision each retreating surge seemed to restore things exactly as they
were, and yet, if I could see all the hidden activities of my kingdom, I would know that every
swell, each minute and each hour, must cause a
thousand thousand tragedies—exposing to hostile, alien eyes hidden weakness and camouflaged
defenselessness. Not a moment passes but somewhere a color secret is exploded, an inedible bluff
called, for even a fish's memory can span ten feet
and two seconds, with hunger as the stimulus.

As the buffalo herons and cowbirds and black

FIG. 42.—RUTH ROSE DIVING AT COCOS IN FIFTEEN FEET
OF WATER.

FIG. 43.—BLACK-BARRED SURGEONFISH, *Teuthis triostegus* (LINNÉ).

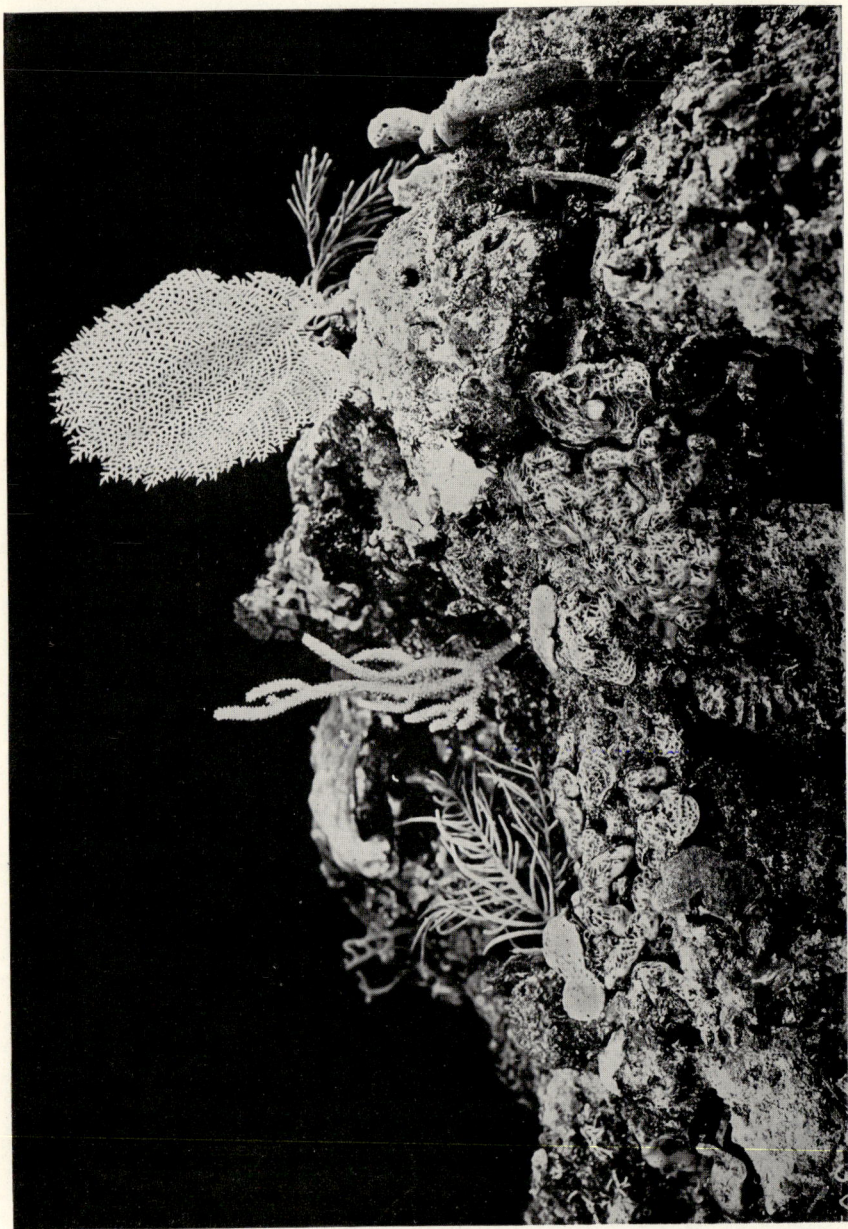

Fig. 44.—A Large Piece of Coral, Supporting a Host of Brilliantly Colored Seafans, Sponges, Shells and other Organisms.

From Saba Bank, 40 Fathoms.

cuckoos gather about grazing cattle, to snatch the disturbed grasshoppers, so, on the outskirts of the surgeon herd, small wrasse persons and others frisked about, darting in to seize some crab or shrimp which the scraping teeth of the grazing fish had dispossessed. Again, as in grazing herds of antelope in Africa, a zebra will now and then be found, so here, mingled in the depths of the three hundred odd, I saw several white-striped angel-fish and as many of my old friends, the yellow-tailed *Xesurus* (Fig. 45).

Fish such as these I take as types of my Grazers —Coral Grazers in particular. My study of the yellow-tailed surgeons applies, with slight changes, to the others—fish which swim slowly about, often in large schools, usually at a low level near the coral or rocks. They are apparently well protected by the poisonous spines on various parts of the body and show no fear of other fish. They may be somber in general body color but they always have some conspicuous mark or patch of color, such as the yellow tail of *Xesurus,* the white tail of aliala, and the black bands in another surgeon (Fig. 43). But however they differ in size, color or sociability, they have one thing in common—their teeth.

One glance at the mouth of a lion, a horse or a rabbit tells us much of their ways of life and their food, and no one could ever mistake the teeth of a surgeonfish for those of a shark or even a snapper. My Grazers, judged by their teeth, fell into four general types, the Hands-and-feet, the Chisel or Horse-toothed, the Stockades and the Parrots.

The first I have described in the chapter on *Xesurus,* the six and thirty little soles sticking up all on edge, and in *Hepatus,* another surgeon, the teeth are absurdly like hands, palms out, with the fingers held close together.   This fish is content with algæ, as I have never found a crab or other marine animal in its food.   The triggerfish are armed with the dental chisels, *Melichthys,* the beautiful black trigger, and the solitary, and preternaturally solemn *Pachygnathus.*   The front view of these fish presents a horrible horse-like appearance, a horse whose teeth are too prominent and too many.

The Stockades are a strange group, with teeth which far excel any instrument of human manufacture.   Details are for the ichthyologist, but consider for a moment the moorish idol and the angelfish.   The astonishingly beautiful white-striped angelfish has a solid outside row of stockade teeth, growing out of a thick, bony jaw.   Back of the front row are four or five layers of teeth, appearing above the jaw in short lengths, looking, on the whole, like a strip of ticker tape or pianola music.   But *Pomacanthus zonipectus,* or more trippingly on the tongue, orange-finned butterfly-fish, has the most bizarre mouthful of any of my Grazers.   At first sight it seems to have a few, large, curiously ribbed teeth, but on closer inspection these are seen to be composed of many, fine, slender individual teeth, like glass splinters.

With all the fish Grazers the price of such definite, abundant, non-motile food seems to be a stiff-

ening of the whole body, activity being superflu-
ous; progression being by fins in place of any undulation, a rolling of the eyes in place of a twist-
able neck and body. But one does not need to go
in enormous schools, hordes or herds, such as surgeons, locusts and antelope affect. The idols keep
in pairs, and they swim and bank, turn and feed
with such unanimity that they might well be a
single moorish idol and its shadow. They have an
outrageous pout which reminds me of a lawyer
friend, and which must be most useful in a Grazer,
since one can graze and yet see upon what one is
grazing without shoving back from the table (Fig.
22). A word as to the Parrot-mouths, which character indeed, has given their name to some of them.
Here, as for example in the puffers, the teeth are
wholly consolidated to form great cutting plates,
usually divided in a hare-lip fashion into four. In
other fish, as the stonewall perch, as the Japanese
call it, the components of the beak are faintly visible, although solidly ossified, the tips showing as
rounded, flat nodules.

When I see what a considerable proportion of my
subjects keep life within their bodies by scraping
rocks and coral clear of the encrusting algæ,
worms, shells, crabs and other growths and organisms, I marvel that the exposed surfaces are not
all as close-cropped as a sheep meadow. But the
clipping seems to hasten renewed growth, and as
there is never any trouble about irrigation, there is
a never-ending supply. Again we must remember
that, strictly speaking, all the fish of this group

also belong to the tribe of Nomads, in the sense that their home is where they are and where food abounds. From my studies of the Grazers it seems to me that they must sleep in some manner inexplicable to us, confining to some one part of the brain the automatic, temporary regulation of eyes and fins. In an aquarium I have never been able to surprise one of the Grazers off guard. The one exception to all this is the triggerfish, and these are chronic loungers and dozers. They lean over anything handy or slant back against a corner at night and are not easily disturbed, although the lidless, staring eyes are never veiled.

As with any community, the more I studied my kingdom the more complex became the various sects and guilds. I could keep on for many pages without beginning to exhaust even my superficial knowledge of the Grazers, for turning to the great patches of sand without my mushroom city, I perceived castes of sand Shovellers and Sifters, to say nothing of sand Waiters who disguised their deadly aggressiveness beneath a thin covering of this white dust,—dust which I could never think of as wet, because as usual there was nothing dry for comparison. To the more gentle sand folk belonged the shovel-nosed rays, the mullets with their delicate chin feelers like the tapping stick of a blind man, and *Polynemis* of the long thread fingers, forever stretching out for knowledge of sustenance. These too, were sand colored, but probably rather as a protective cloak against the peering eyes of enemy Nomads.

The great backbone of my population, the host of "common peepul" was what I called Percolators —although some of them were aristocrats and many did not percolate. As a whole, however, they lived their lives in and out of the coral and rocks, never becoming surface lovers, nor settling down in any special crevice. Still they were local optioners in point of residence, and many a time I recognized the same individuals in the same coral palace grounds. In taste they were omnivorous or carnivorous, seldom wholly vegetarians and never strictly grazers. Like New Yorkers at lunch hour they were victims of idle curiosity, and I shall never see a throng watching with breathless interest the working excavators, or the rhythmical riveters on some new building, without remembering the crowd of small Percolators who always rushed toward me when I first submerged, swimming rapidly with a My-Word!-see-what's-here expression.

My percolators belonged to many families and systematic gens, and their diversity of habits within the limits I have set would fill volumes. The most abundant was probably the beautiful blue-lined golden snapper, *Evoplites viridis,* which is one of the most beautiful of fish. This may be taken as typical of the group. In a dozen stomachs I found that crabs and very tiny fish each occurred five times, shrimp thrice and snails once. They have little social instinct and while a score or two would gather quickly at hint of a repast, yet they were never closely associated in schools.

They ran or swam with many of the other

medium and small Percolators, the most brilliant
being wrasse, with unbelievably harsh and gorgeous
pigments and patterns. *Thallasoma,* or the
pousse-café fish, with its purple and yellow and
green stripes, always formed a large percentage
of the crowd of fish whirling about my hand and
helmet when I held a bit of crab as lure. *Dermatolepis,* the big, high-backed, golden-spotted sea
bass, must be considered as giant Percolators, for
they were always trying to push through crevices
and archways too small for them. They were ugly-
natured as well as bold, and needed only a little
more courage to attempt to ham-string me when
I was not looking. As they became angry or over-
excited they showed their spleen or nervousness
by changing color, thinking nothing of shifting
from white to black, always with the yellow gold
spangles shining clear. At the other end of percolate size were the tiny *Runulus,* midget eelet
wrasses, who eddied in and out between my fingers
without my ever succeeding even in touching them.
When I tell that within a few minutes after taking my coral throne, I often had five hundred Percolators swimming close about me, the vast numbers of my subjects (if not their loyalty other than
gastronomic) may be realized.

Two more castes remain, the Squatters and the
Villagers—my favorites of all my fish. They were
of greatest interest when compared with one another, for while the latter had individual crevice
homes, yet they were built along normal fish lines,
while the Squatters, although they might spend

only a short time in any given spot, were all physically adapted and modified for life in and over and around rocks and coral.

The most abundant of the Villagers were the brown Pomacentrids or demoiselles. They were everywhere and yet each one had its own little domicile—a hole, or crack or crevice where it resided and which it defended against all comers. A sight of which I never wearied was to see a big *Xesurus,* if not indeed a grouper itself, come barging slowly along, when suddenly out from the very coral rock in its path there would shoot a diminutive demoiselle, fins erect in righteous wrath, and actually rush at the offending giant. The gesture of home defense was so real that the attacked one, if a small fish, usually turned tail and fled at once, or, if the dignity of size had to be maintained, the surgeon or grouper would veer slightly to one side, as if recognizing and acknowledging the excellent motive of irritation, but saving its own face.

All my life I have had a weakness for gobies and blennies, and now that I was able to sit upon a rock and have them come out like elves and gnomes, and skip and slither about at my elbow, my fondness grew to real affection. Of all fish these give the impression of being less completely bound up in fishiness. I am sure that they would make splendid pets, and would do all they could to cross the border line which divides the inhabitants of the realm of water from us elemental mongrels. We, lords of creation indeed! who must needs breathe one thin medium, support ourselves upon a thicker one, and yet, although our body is two-thirds of another, perish miserably when immersed in it.

I have given above a few stray notes of a minute fraction of my shallow water kingdom, observed in a succession of fleeting moments of time. Its chief value is to show our ignorance of this cosmos—and to stimulate at least my own desire to go and learn more.

When I reached Cocos Island I had with me a list of thirty-eight species of shore fishes which had been collected twenty-six years ago. Of these I was able to secure and identify twenty-three, in addition to fifty-seven others which had never before been recorded from this lonely island. During my stay therefore I observed eighty species, making a total of ninety-five altogether. I have notes on at least a third again as many which were too wary for me, although some of them would swim up to the very glass of my helmet and gaze impudently in at me, and which were quite new species. The details of this fauna, their names and relationships, colors and food belong elsewhere, but I desired here to give a shadowy hint of the mode of life and the personalities of a few, in the pitifully inadequate method of adulteration through the medium of human thought and words.

CHAPTER XIII

SEVENTY-FOUR: AN ISLAND OF WATER

My sub-title is not a mere meaningless catch phrase, but a reality. In Ruth Rose's chapter on Osborn Island she concerned herself, and rightly, not only with the things bound to earth, but the birds flying overhead and the sea-lions on the beach who live their active lives beneath the waves. The island of which I write is a tiny speck of the bottom of the Pacific Ocean and my interest in this has to do with both this bottom land and its inhabitants, as well as with the host of creatures which swims and floats to and fro over it, at various elevations, up to the surface itself.

I justify my title in another way. The dictionary defines island as a body of land entirely surrounded by water, to which characterization my island has the more logical right, for mine uses the word surround in the completer sense of being covered as well as margined by water. Etymology even comes to my aid, in the old Anglo Saxon *ea-land,* which may be interpreted *water-land* or *sea-land.* This is exactly what I established in mid-ocean.

As to my title itself, taken from the number of this station, no defense is required. Is it not the most holy and lucky of numbers, containing the

317

Hebraic significance of all that is abundant, satisfactory and complete! And as for precedent I
can indicate in olden times "The Seventy" the title
of the seventy-two translators of the Septuagint,
and (it seems only yesterday), who of us will forget, who have seen and heard them,—the thousands
upon thousands of long, slender stems, with upraised muzzles alert and ready like the fangs of
faithful watch-dogs, stretching on and on in an
unending, unbroken, unbreakable line, over hills
and through valleys, like the towers of the Great
Wall—*les soixante-quinze!*

My intention in regard to an island of water was
simultaneous with my turning from the jungle to
the ocean—exemplifying my passion for small, restricted things. In many ways an island is much
more significant than a continent, a solitary tree
than a jungle, the life history of a single family of
living creatures, or of one species, or, better still,
individual, than casual studies of an entire phylum.
This accounts for my biased researches in times
past.[1] I fear that the same characteristic would
always rob a jail of its horror—there are reasons
why I would rather be the Prisoner of Chillon
than the Wandering Jew.

When I began studying the oceanographic voy-

---

[1] *Natural History of Pheasants,* N. Y. Zool. Soc.
"Four Square Feet of Jungle," *Zoologica,* II, p. 107.
"The Bird of the Wine-Colored Egg," *Jungle Days,* p. 182.
"Birds of a Single Tree," *Zoologica,* II, p. 55.
"A Jungle Labor Union," *Edge of the Jungle,* p. 149.
"A Chain of Jungle Life," *Jungle Days,* p. 3.
"The Three-toed Sloth," *Zoologica,* VII, p. 1.

ages of past years, one thing stood out at once—
the tremendous distances covered. The ship would
stop to sound, make a haul, and then up sails or
steam, and away a few hundred miles to the next
station,—the very name station being significant of
railway speed. This was necessary, for pioneers
in any field must be peripatetic. Much good
Columbus would have done the world, milling
around in one spot in mid-ocean, or Balboa if he
had been content to rest at the foot of his Darien
peak. There is still need for hundreds of more
voyages of widest range before we can know the
distribution of ocean life with any accuracy.

My objects in the *Arcturus* adventure militated
against any prolonged study of a single locality.
To learn anything of the Sargasso Sea and the
Humboldt Current I must cover hundreds and
thousands of miles, and this I had done. But away
at the back of my mind was an obstinate intention
to have a try at making an island out of an enormously tall column of water resting on a limited
bit of very wet land. I was conservative in my first
attempt and decided to select a place where the
pillar of water was less than a mile in height. I
say *height* advisedly, for if anything is worth
studying intensively, one must absolutely identify
oneself with it. Some of the greatest joys of my
life come when I shed the unlovely man-body thing
which I am condemned to carry about through life
as transportation and periscope to my mind and
soul. For the time being I must become pheasant,
protozoan, sloth or tree.

Now I was to become, not only a fish, but one on the bottom—on the face of my island, so that I must speak of the height, not the depth, of the water overhead. It is an easy thing to do, if you love to do it, and on land the reverse is equally facile, for the depth of air over a given place becomes almost a trite term, when you have flown over it a score of times.

I cheated perhaps a little about my water island, but I was so anxious to have it a success, that I was willing to load the dice a bit. By this I mean that I let myself be influenced in choosing the spot by the memory of an unusually splendid haul which I had made not far away a few weeks before— not a very heinous thing to be sure, but not quite as sporting as would have been steaming blindly ahead and suddenly stopping anywhere in open ocean.

When I came to think of all the details of my new endeavor, the subconscious worry and fear of the whole expeditionary responsibility, which was always hanging over me, became more vivid—floating to the surface of my mind and unpityingly pointing out the situation. A ship is made to travel, its engines to throb, and although I was in complete command, yet the shadow of my old passenger subordination always lay heavy upon my decisions. There seemed too, something against all the traditions of the sea in thus wilfully turning a perfectly good vessel into a derelict of sorts, even for a time. I pictured the weed and barnacles on the keel as sprouting forth in awful rapidity of growth during the period of inaction, the engines becoming rusty, the engineers and oilers falling asleep one by one—indeed before I knew it I had visualized another Flying Dutchman, only under a static instead of a dynamic spell; I seemed to be laying the foundation of a Pacific sea of dead ships.

I prepared for the experiment by the study of a wholly different type of fish fauna—the shore fishes of Cocos Island,—that speck of land so beloved by the pirates of old, about five hundred miles off the coast of Panama. If preliminary success was augury of good luck, I should have been contented, for the finny inhabitants of Chatham Bay yielded up their secrets in wonderful fashion. The rainy season had been a jest at the Galápagos, but no season ever merited it more than at Cocos. As I spent most of my time in my diving helmet beneath the surface I hardly noticed the constant downpour, but it was a fact that the air was saturated most of the time. Dwight Franklin one day laid a water-color sketch marked "Cocos" on my laboratory desk, a composition consisting of a wide expanse of sea, with a small smudge of a rain storm in the center; a joke but not an exaggeration.

When I had once halted my ship in mid-ocean I had no hesitation in knowing what to do. I wanted to learn all I could of what flew in the air, floated on the surface, dived in the depths or burrowed into the substance of this tiny pin-point in the great Pacific. But now that I have finished and steamed away, and weeks have passed since the last dredge came up, I am confused as to the manner of telling about it. What I did day and night, of dredging and trawling, was done so blindly, so gropingly, what came up was such a pitiful fraction of the great mass of life which must be below, that I feel like a deaf, dumb and blind person attempting to interpret a wholly new and strange world.

With more usual islands, one naturally begins with the life of the ground, then that of the trees, and finally with net and gun and glasses one collects and studies the beings of the free air. Here I shall reverse the process and begin with the top of the water column.

On Sunday, May twenty-fourth, in the late afternoon we pulled up anchor at Cocos Island, and steamed westward out of Chatham Bay, slowly encircling the island. After skirting the southern headlands and passing the zone of uncharted shore, I gave orders to turn south, and in a swirl of wind and rain Cocos changed from dull green to grey, and finally was lost in the black mist of night. Under slow speed we crept southward, and at dawn, with the mountainous little island just visible on the northern horizon, Bill Merriam let go the sounding weight. Minute after minute the piano wire hummed its song of swift descent into the blue waters, and came to rest at last when the seventy-five pounds of oval iron weight struck bottom in seven hundred and seventy-one fathoms —both weight and depth sonorously reiterating the sound of the new station's number—*Seventy-four*. Thus was made first contact with my island of water.

For the next ten days, from early on this Monday of May twenty-fifth, to five in the afternoon of June third, we floated, within as small an area as was possible without anchoring, above the isle of our own making. I will give it the dignity of a definition such as used to be printed in our school geographies:

The center of the island is four and a half degrees of latitude north of the equator, and eighty-seven degrees of longitude west of Greenwich. Its nearest terrestrial neighbor is Cocos Island, which is due north, one degree, or sixty miles. To the south-west, three hundred and fifteen miles away, is Tower Island, the nearest of the Galápagos, and the nearest point on the American continent is Florena Point, Costa Rica, three hundred and five miles northeast. The inhabitants of *Seventy-four* are engaged chiefly in fishing, its exports being fish, sea-cucumbers, jelly-fish and other marine products, while its imports consist of entangled dredges, coal ashes and fresh-water rain. For ten days it was a colonial possession of the United States. It has now reverted to No Man's Land and the realm of memory and imagination.

Cocos vanished from sight early in the evening of that damp Sunday, yet day and night thereafter we were constantly to feel her influence, even when sixty miles away. The rain steadied to a downpour, and as I looked out of my cabin door, the deck was a maze of starred splashes, and the edge of the blackness a thin screen of slanted, pearl-grey lines etched on the substance of night.

At midnight the unending warp of rain still threaded the invisible sky and sea. I lay in my bunk and listened to the unearthly cries of the confused sea-birds. The high, shrill, pitiful notes filtered through the murk, and then, suddenly, several ghostly forms would shape themselves, fluttering tremulously far out in the driving wind and rain, proving that the darkness was not darkness after all.

In the museum of Uyeno Park, Tokio, there was once an incomparable collection of kakemonos —the rarest work of the best old masters of Yeddo and China—all taken now by the earthquake. Unknown to me, there was hidden deep within a forgotten cell of memory, a clear-cut vision of one, showing sea-gulls flying in the rain. And now, on this rainy midnight at sea, the picture flashed to consciousness, for there before me, framed in the long rectangle of my cabin door, Hokasai's kakemono lived again.

I lay back in the bunk, writing on my drawn-up knees, my posture recalling Stevenson or Twain in everything except the value of what I wrote. A half hour passed and the rain was Monday's rain, when I heard a gentle whipping of wings— the sharper tone which is given out when wings are very wet. In mid-air in my cabin, beating a little cross current to my electric fan, was one of the fairy terns of Cocos. As I looked, the immaculate little beauty fluttered upward and poised close to the wall light, then sank slowly and came to rest on my knee. I finished my sentence and began to write a description of the dainty bird, while it ruffled and shook and settled its plumage into place, showering me with drops. I felt no envy of Stevenson or Twain now. For the space of several minutes we looked at each other, the tern much the more composed and less breathless of the two. Then, lightly as thistle-down, it rose, fluttered over to my desk and alighted in the middle of a large map of Cocos Island which happened to be lying there (Go ahead, Reader, say it yourself, I won't bother to write it!).

For a long time the bird preened its white plumage, looking about with its dark, quick eyes and burying the slender beak deep in the feathers, fluffing them out. The chicory blue of the beak was just the touch needed to set off the snow-white plumage. As it preened, it walked slowly about on the paper Cocos, the violet blue webs between the toes pattering softly. Then the long, angled, capable wings were stretched, high, high up, and a half dozen quick beats lifted the whole little being, making palpable the thin air. Without haste, yet without hesitation, the fairy tern drifted out of the door, glimmered like a painted kakemono ghost for a moment, and vanished. I watched the same slanting lines, listened in vain for any last call it might have sent back, and wondered whether I had not dreamed a dream. But the map of Cocos Island showed a cluster of little, swollen blisters where the damp drops had raised the paper, and to the paper-flat slopes of Mount Harrison there clung a tiny feather—not soft and downy from the body, but a little tertiary from the wing itself.   Again I looked out and marvelled how such a pinch of a white fluff of a bird, scarcely a foot in length, weighing less than five ounces, could have the courage on such a night to leave light and shelter and safety—for it had showed not the slightest fear of me—and launch out into the driving rain, with the nearest tree sixty miles away.

During this first night of rain and wind, boobies by the dozen also sought haven on the lighted steamer, after a fashion far otherwise than the white tern.   They heralded their coming with squawks, sounding muffled through the distance and rain, and then flopped to the decks or against the cabins with a bang.   Thereupon they raised their voices to the highest pitch of raucous outcry, launching awful protests, screaming curses of anger and fright until the steamer rang with the noise.   Toward morning a great red-footed booby bludgeoned into my room, missed my face by a narrow margin and thrashed his way out again. I snapped on the light and envisaged a mill of devil birds.   At my threshold my visitor encountered another of his kind, a hated rival of long standing, it appeared to me.   In addition each immediately credited the other with all the blame for the storm, the confusion and an intense dislike for this new-found sanctuary.   A battle ensued, and with beaks gripped on one another's persons, the combatants remained locked, lying on their sides, squawking full steam through half-closed beaks until I went out and hurled them both over the rail. After the voluntary leave-taking of the white tern I had no fear for the safety of these great birds, provided the plunge cooled their frenzy of hate.

The rain ceased just before dawn and gave place to a strange, hard sunrise—a scarlet slit in the ash grey of the east, and an unreal, pallid, greenish expanse in the north. In this eerie light, at five-thirty, we made the first sounding which I have described.

In the ten days during which I floated over my island, I had rather remarkable luck in recording birds. I observed seventy-four altogether, comprising thirteen species. Six of these were sea-birds from Cocos, which had come this great distance to some favorite feeding ground, or in a few instances had perhaps been blown farther than they had intended to fly. Of those which came on board in nights of stress and storm, some were obviously exhausted but most were apparently strong on the wing, and only confused and distracted from their true course by the sudden vision of the ship's lights.

Five other species, three petrels and two shearwaters, were true pelagic birds, feeding as they flew and paying no attention to the vessel. Then there were two strays, probably storm driven, a gull and a warbler.

To be more specific, one day a frigatebird flew past with its marvellously slow wing beats, headed for Cocos. It may have been out for days without tiring, and in the case of such low-lying storms as those hereabouts, could easily rise above the level of the rain. The two Cocos boobies, the red-footed and the white-breasted, came in numbers to our lights. These birds travel thirty and forty miles to and from certain fishing grounds, but are not capable of nearly as prolonged flight as the frigates. The boobies of Tower Island feed for the most part, forty miles away from home in the direction of Indefatigable, although fish seem quite as abundant near at hand, and here at Cocos the same inexplicable habit would seem to hold. We caught several boobies on the decks and caged them for exhibition in the Zoological Park. When first caught they were fiends incarnate, dashing themselves against the wire, screaming and striking fiercely with their powerful beaks. Within three days they had become quiet, almost gentle, making no attempt to injure the hand which provided them with fish. A hint of the wonderful sight and balance which they use in diving after their prey is shown in the way they catch pieces of fish, for no matter how swiftly it is thrown or at what awkward angle, with a slight twist of the neck the fish is caught.

Shearwaters were in sight almost every day, the dusky, and the larger, white-fronted species. One day while watching a school of tunnies leaping high in air, a dusky shearwater wheeled into sight directly in front of the bow. I watched it with the glasses for a time and, as I had paper and

Fig. 45.—White-striped Angelfish, *Holocanthus passer*, Valen.

Fig. 46.—White-lined Triggerfish, *Melichthys bispinosus*, Gilbert.

SHORE FISH OF COCOS ISLAND.

Fig. 47.—A Blind Deep Sea Fish, *Bathypteröis* sp., whose Chief Contact with Life is by Means of the Long Tentacle-like Rays of the Pectoral Fins.

pencil, I followed its flight. I know of no bird better named than this. First on one side of the bow, then the other, the bird described loops, doubling almost into figures-of-eight. At one point in its course, it put on full brakes with wings and feet, spattered for a few feet through the water, with quick paddling webs, snatched a small fish, swallowed it and left.

When I had it in the field of my glasses I saw what, to me, was a wholly new observation—the dipping of the under wing-tip well into the water at almost every outer edge of the turns, and not only this, but a very apparent throbbing or successive fluttering of that wing alone (the other being held quite still), as if to increase the braking power, or the fulcrum value of the heavier medium. It reminded me somewhat of my old days of pole-vaulting, when, running at full speed, I struck the tip of the pole into the ground. Time after time I watched the little furrow which the wing made, and saw the tremulous pressing against the slight hold of the water. After forty or fifty observed repetitions, I have not the slightest doubt of the material assistance which this habit gives to the ease of swift pivoting and steep banking.

Mother Cary's chickens or stormy petrels were present on most days, regardless of waves and winds, flickering cheerfully about their business of finding small prey. Leaches and dark-rumped petrels I expected to find, but when a white-faced petrel (*Pelagodroma marina*) flew on board late one evening, I knew I had a prize. This bird has its center of distribution near Australia and New Zealand, but here was a straggler thousands of miles away from home, and yet strong on the wing and in good health. It became confused by the ship's lights, flew on board and was not able to rise from the flat deck. It is accidents such as this which keep scientists from becoming conceited, realizing as they must, how much of their knowledge depends on chances.

The stray gull was peculiar to the Galápagos, and it flew around the ship wing-wearily one morning, like the one I had seen the week before at Cocos. Storm or wind or some strange wandering instinct must have brought them over more than three hundred miles of ocean. The white tern and the two species of noddies were all Cocos birds, out fishing when the drenching rain and high wind forced them to come aboard for rest. Numbers of birds must perish in every severe storm, for although these seabirds have well-oiled plumage and webbed feet, yet a strange fear of the water obsesses them, and they alight on its surface only as a last resort, dreading some danger unknown to me, whether of some dangerous fish, or of the fatal water-soaking of already drenched wings. There remains of my island avifauna only the most unexpected visitor—a dainty, Cocos Island, yellow warbler, which appeared one morning in the rigging. The wind of the preceding night had blown from the east, it was not over strong, and the night, although dark, was without rain, so the arrival of this land bird was wholly unexpected. For an hour it preened its plumage, then half-heartedly sang a single phrase of its simple ditty. It next flew down to the deck where, with the skill of a professional flycatcher far transcending that of an ordinary warbler, it caught two flies which were humming about a dead fish. A moment later it rose, and in a steep ascending spiral, after gaining an elevation of about two hundred feet, it darted along the compass line for Cocos, fifty-eight miles away.

As to claiming completeness of representation of vertebrate classes on my island, I announce failure at once. No amphibian, whether frog, toad or polywog existed nearer than the American mainland, but this was the only group missing. I lay flat in my bow pulpit one day, while we were slowly steaming in a great circle, drawing a half dozen large tow nets, when I saw two rocks ahead, just awash. Before the first impression could crystallize into actual belief, I detected the rounded, upturned heads, and knew that the class of reptiles could be included in my island fauna. They were big, green sea turtles, although one belied its name for its shell was a warm brick red in color, dotted here and there with large, white barnacles. They drifted slowly past me, one on each side of the *Arcturus,* merely turning their big heads, but not moving otherwise until they were tumbled by the bow waves, when they immediately dived.

Two species of sea mammals paid the *Arcturus* and the island a visit within the ten days' space;

three great schools of dolphins churning past,
headed northeastward, while on three other days
a school or sound of small whales, some species of
blackfish, passed, going in the same direction. The
third lot, twenty-seven in number, appeared in the
late afternoon of our last day. They split up
temporarily, twelve or fifteen coming close to have
a look at this strange, larger whale. They rolled
ponderously about, sighed audibly with sprayfuls,
and steamed steadily after their fellows.

Although my island is sixty miles south of
Cocos, yet now and then I find a dead land insect
or some seeds in the surface towing nets—a tiny
cockchafer or June bug, a water-worn hawkmoth
and a flying ant. On May 29th twenty or more
dragonflies appeared suddenly on board, hawked
about, catching nothing that I could see, although
since the warbler had taken the lonely pair of
flies, I had seen about a dozen others on board. I
caught one of the dragonflies and found it was a
large species peculiar to Cocos, with wings hyaline
except for a black spot near the base of the hinder
pair. On another day a butterfly flew about the
ship for hours, one of the strong-winged, leaf-
shaped, orange and black brassolids common on
the island to the north.

All this radiation of living creatures, birds and
insects, and, as we shall see, plants and fish, over
half a hundred miles from a small island, across,
rather than with, the prevailing winds and currents, gave me an entirely new idea as to the effectiveness of oceanic distribution, and one which was

rather destructive to former theories I have held. If my island had suddenly appeared above the surface, and if we granted a certain amount of scientific license in the matter of soil ready to hand, there would have accrued to it a surprising number of living beings, judging even by the restricted space observation from the deck of my vessel and from the brief time period of ten days.

This point of view is thrilling to me, and some day, when my physical activities become curtailed by age, so that I shall be compelled to shift from tennis to golf, from dancing to contemplation, then I will give up active exploration and diving and hunting, and settle down upon a barren desert island. If one recently elevated by a submarine earthquake or other terrestrial disturbance is not available, I shall manufacture one myself out of concrete or coral and sterilized earth, off some interesting shore or bank of river, and day by day, I shall watch the accidental populating—the simple beginnings of the struggle for existence between seed and seed, animal and animal. Then perhaps I shall see a little more clearly into the meanings of the apparent terrible confusions already in full swing, which in great jungles so cobweb my brain and mind.

The possibilities which might result from the ten days' emigration to my island, supposedly recently emerged, are as follows (I have allowed myself the liberty of considering that the three drowned insects are still alive) ;

CLASS I PLANTS

2 cocoanuts
3 other plants

CLASS II ANIMALS WHICH COULD BECOME ESTABLISHED AND BREED

2 species of boobies
3 species of terns
8 species of shore fish
4 species of shore crabs
1 species of fly
2 species of feather fly
1 species of dragonfly
1 species of ant (female ready to lay eggs)
1 species of hawkmoth (female ready to lay eggs)
1 species of June bug (female: her eggs were very small, but having resurrected her, I crave indulgence to imagine this beetle's eggs as ready to hatch into grubs)
1 species of butterfly (? This is included here only by the courtesy of ignorance, for this insect was seen, not examined, and so may not have been a fertilized female)

CLASS III EXISTENCE UPON ISLAND LIMITED TO LIFE OF THE INDIVIDUAL

frigatebird
gull
yellow warbler

CLASS IV PELAGIC SPECIES WHICH MIGHT COME ASHORE AND BREED

3 species of petrel
2 species of shearwaters
1 species of sea turtle

It is amusing to follow in imagination the direct possibilities of the first stocking of the island.

Class I—Plants—Both cocoanuts were living and one already sprouted. They were floating buoyantly, and had apparently only recently been immersed, as there was no hint of algal or barnacle growth. Two of the plants were growing on a floating log, and the third—a long section of coarse creeping grass—was floating by itself. The grass and one of the other plants, which might have developed into a shrubby growth, both sprouted at once when put into soil in a deck garden box.

Class II—Animals capable of establishment and breeding—The fifty-odd individual terns and boobies of five species, would have found a recently emerged island a perfectly satisfactory home, with an abundance of sea-food, and rocks and crevices for their nests and eggs. The dozen kinds of shore fish and crabs which I obtained from floating logs would experience no radical change and find plenty of food in shifting their shelter from logs to rocky shallows along shore.

Among the most important members of the new fauna would be the dozen flies of presumably both sexes which were on board. Dead and decaying sea creatures would immediately furnish them with provision for their eggs or maggots, and even if most of their number were devoured by the dragonflies and the yellow warbler, their race would probably be preserved, furnishing satisfaction to the small fraction of their descendants who lived, and food for many other creatures.

On the boobies there were hosts of feather flies, so many in some cases, that they flew off at the slightest disturbance, and could thus be counted upon as another source of food. The ant and the hawkmoth being both females with eggs almost ready to be deposited, they might very reasonably already be fertilized. The possibility of these particular plants being the kind upon which the caterpillars of the moth would thrive, are slight, yet the thousand and first chance has many times insured the life of a whole race. The queen ant would not have a very difficult time in establishing a colony, but the grubs of the June beetle would be lucky indeed if they found sufficient nourishment in the newly grown roots available in this instance. The dragonflies would need some rain pools, and out of the score, a pair or two might survive and propagate their kind, their food consisting of what flies they could capture, with the possibility at the last of devouring one another. Finally, I have included the butterfly in this class, not because I succeeded in catching and examining it, but on the chance that, like the moth and the ant, it might possibly be a gravid female.

Class III—Creatures doomed for the present to mere existence on the island—frigatebird, Galà-

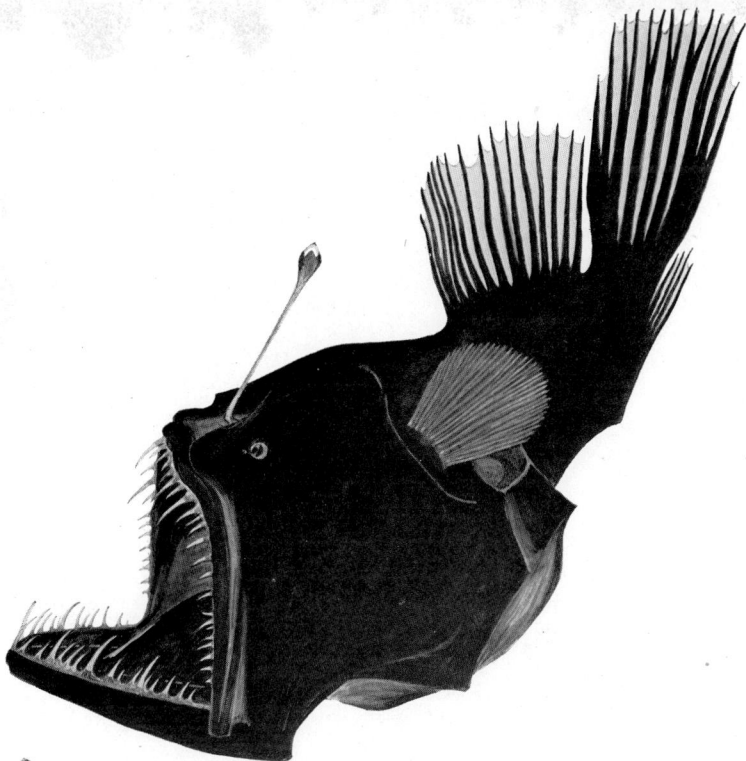

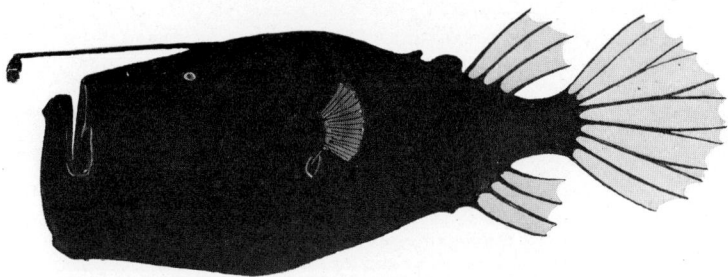

FIGS. 48, 49, 50.—THREE NEW SPECIES OF LANTERN-BEARING SEA DEVILS FROM STATION 74.

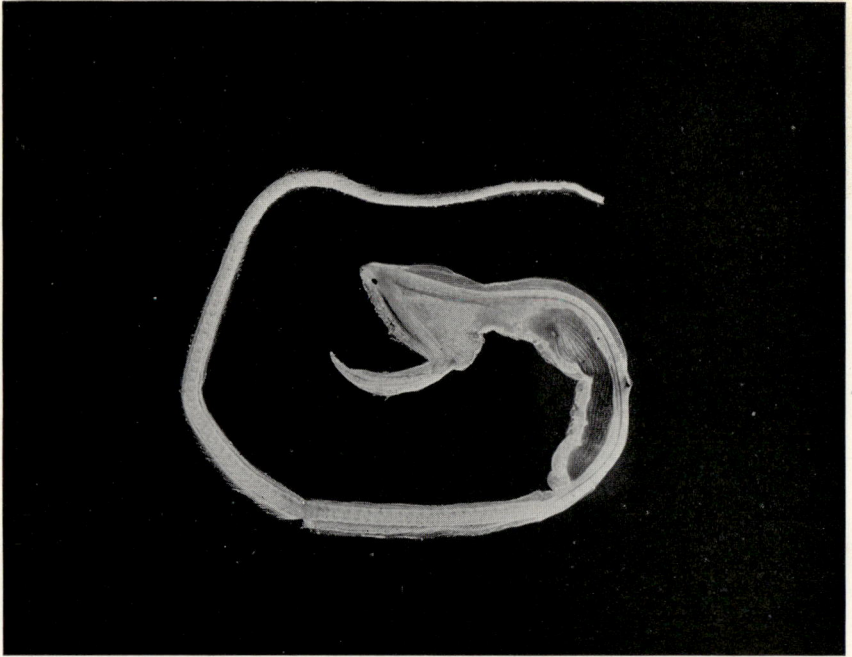

FIG. 51.—PELICAN-FISH, *Saccopharynx* sp.

A new species, white in color, dredged at a depth of five hundred fathoms.

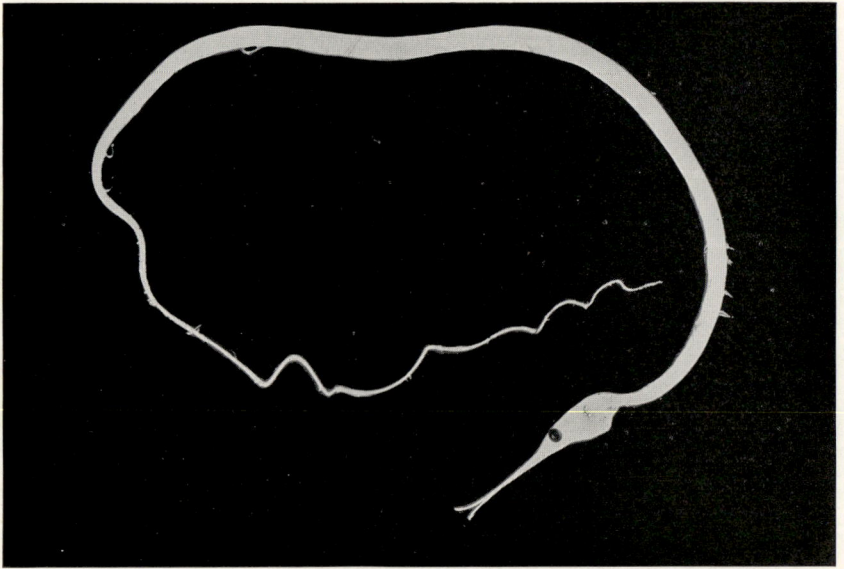

FIG. 52.—A SILVERY SNIPE-EEL, *Nemichthys* sp.

From a half mile depth.

pagos gull and yellow warbler. With only a single
sex, permanent establishment and increase would
be impossible, but with sea-food for the first two,
and flies for the warbler, all would survive for the
duration of their lives. In fact at Cocos, I saw a
number of mockingbirds and yellow warblers
feeding exclusively along the line of tide, picking
up tiny shrimps and other forms of marine life.
So my island warbler while waiting for the forlorn hope of an arriving mate, would not have to
depend upon the precarious diet of flies, which
might have succumbed to the attacks of dragonflies, or to some more subtle inimical agency.

Class IV—Pelagic species—This division is
merely to visualize the possibilities of three species
of petrels, the two shearwaters and the sea turtle
coming ashore and making their homes on
Seventy-Four, as they could on any oceanic island which provided crevices of rocks and, for the
chelonian, a sandy beach.

At the expense of being statistical, but in order
to sum up the possible surface and aerial inhabitants of my imaginary raised island, I present a
census of the complete initial population.

VEGETABLE KINGDOM

4 species of living plants of 5 individuals

ANIMAL KINGDOM

1 species of fly	12	individuals
2 species of feather flies	299	"
1 species of dragonfly	20	"
1 species of ant	1	individual

1 species of hawkmoth	1 individual
1 species of butterfly	1 "
1 species of beetle	1 "
4 species of shore crabs	25 individuals
8 species of shore fish	40 "
1 species of sea turtle	2 "
2 species of boobies	30 "
1 species of frigatebird	1 individual
3 species of petrels	35 individuals
2 species of shearwaters	18 "
3 species of terns	23 "
1 species of gull	1 individual
1 species of warbler	1 "
34 species	412 individuals

So even with the scattered and imperfect observations which I was able to make, I could see my island stocked with plants and insects, shore fish, crabs, sea turtles and birds, the whole numbering over four hundred individuals.

The earth is altering with eldritch rapidity before the onrush of increasing numbers and the destructiveness of mankind. The details of early evolution and the clarity of primitive relationships are daily becoming less distinct, more complex. It is a wonderful thought that in addition to the continued chance of deciphering the primer of paleontology, and the interpretation of the body and mind of the young of all animals, there is the possibility of learning much from close observation of beginnings, such as this stocking of an island—an island, desert in the very deepest meaning of the word.

CHAPTER XIV

DAVEY JONES' GOBLINS

Not long ago a man named Grahame wrote of a strange creature, "He was sticking half-way out of the cave, and seemed to be enjoying of the cool of the evening in a poetical sort of way. He was as big as four cart-horses, and all covered with shiny scales—deep blue scales at the top of him, shading off to a tender sort o' green below. As he breathed, there was that sort of flicker over his nostrils that you see over our chalk roads on a baking windless day in summer. He had his chin on his paws, and I should say he was meditating about things."

Forty years ago another man named Collett scooped up another equally strange creature from the surface of the sea and wrote of it, "Head enormous; the body slender, compressed, mouth oblique. Spinous dorsal reduced to a single cephalic tentacle, the basal part of which is erect, not procumbent. Teeth in the jaws, on the vomer, and the upper phryngeals. Gill opening exceedingly narrow, situated a little below the root of the pectorals. Soft dorsal and anal very short; ventrals none. Abdominal cavity forming a sac, suspended

339

from the trunk. Skin smooth; a long tentacle on the throat."

Is it not a very sad thing that we must admit that the first description refers to a fairy-story dragon, and the second to a very live fish, first cousin to that in Figure One of this book!

Some time ago when I had read and written scientific facts until my brain whirled, I sought relief one evening by looking at dragon pictures by Parrish and Rackham, and then I became scientific again to the extent of comparing them with colored plates which I had had made of deep sea fishes. To my delight I found that I could duplicate or actually improve upon every character of dragons or gargoyles. After one has become acquainted with the everyday inhabitants—villagers, aristocrats, commoners—living today in the deep sea, Dunsany, Barry, Blackwood, Grimm, Sime— all these lose force as inventors of fairies, hobgoblins and elves, and become mere nature fakers. For in these abyssmal regions there are fish which can outdragon or outmipt any mere figment of the imagination; crustaceans are there to which the gargoyles of Notre Dame, the fiends of Dante's Purgatory appear usual and normal.

I wonder, if at some momentous happening in life everyone does not have the sudden recurrence of an emotion which has not been experienced since early childhood. Mere height or depth never affected me,—I could always look with pleasurable exhilaration over the edge of a precipice or down from a roof. But sometimes under the stars, when there came the realization of cosmic space, or at my first glimpse of moon mountains through a telescope or my first trip in an airplane,—then I shuddered to my soul, and my heart skipped a beat. I remember pulling in a kite with all my might, trembling with terror, for I had sensed the ghastly isolation of that bit of paper aloft in sheer space, and the tug of the string appalled me with the thought of being myself drawn up and up, away from the solid earth. This, my boyhood's very real terror, returned to me one day on the *Arcturus* when we had lowered one of our first deep trawls, and I happened to touch the wire cable extending down into the water of mid-ocean. It hummed and vibrated under my hand, and for a moment stark terror possessed me again as I realized where the net was—a full mile beneath the ship and sunlight, in a region which for power over the human imagination and for utter inaccessibility compares only with interstellar space.

Only once again did I experience this—when, in diving helmet, thirty feet beneath the surface, I was struggling against a bad swell on the steep slope of Tagus Cove. I stumbled and began slowly to slide and drift out and down. I grasped at a bit of seaweed and it broke off, the heaving waters turned me partly around. My foot struck against a coral boulder and a sea-fan gave me solid anchorage. For a minute or more I stared through the glass window—down, down at the terrible translucent blue-blackness of that abyss. There would have been no quick, smashing drop as over a dry precipice in upper air, but a slow, awful rolling, with an unhasting death from cold, pressure and blackness. All this terror was wholly needless, but obvious methods of escape, of safety, were erased for the moment, and my agonized mind was occupied only with dread of this cosmic peril. My reason for all this apparent personal digression is to try, by every means in my power, to make real and vivid to the mind of the reader, the unearthliness of the depths of the sea, and to prepare the background for the strangest backboned animals living on this planet today.

The simile between interstellar space and the ocean depths might be carried to any lengths. Coupled with our inability actually to penetrate either of these regions, we find ourselves of necessity mere peerers in the first instance and blind gropers in the second. In mid-ocean, whether we skim the surface with nets or draw them at a half-mile depth, or drag our dredge slowly over the bottom, the result is a gamble, and may be nothing or the richest of hauls. The merest tyro yachtsman has quite as good a chance of capturing wonderful new creatures as the most experienced oceanographers in the world.

As I have said in the preceding chapter we spent ten days at Station Seventy-four, in mid-Pacific, one degree or sixty miles south of Cocos. Throughout all the time that I was collecting and studying the surface creatures, I was fishing and trawling and dredging deep down—making the most of every piece of apparatus to learn about the inhabitants of this vertical column of water. My success was far beyond my expectations, and comparable only to the results of intensive work in the quarter of a square mile of jungle in British Guiana.

Within the ten days from May 25th to June 3rd I captured one hundred and thirty-six species of fish in this one spot, and at least fifty species of crustaceans.

The thought of conveying in a single chapter any clear conception of the life at varying depths which I discovered even at this single Station, is like trying to reduce the sights and activities of a twenty-ring circus to a single paragraph. That must be left to another entire volume. Here only one thing is possible,—to present a few individual vignettes, each of which will give some dominant idea of deep sea life.

The surface fauna is visible to us from the air and therefore intelligible. Here is warmth and sunlight, and even we ourselves can dive a little way into the water and live. Here are plants and animals, courtships and deaths. The plants grow in the sunlight and the animals feed on the plants, and in course of time die and their bodies begin slowly to sink downward. We can follow them only in imagination, using the knowledge gained by nets and trawls, thermometers and photographic plates. The sunlight gradually loses its power as we sink, the red rays going first—and soon we are in the violet blueness of moonlight. It is cooler, and there is a weight of water which at the surface we never experience. On and on we go until, at a depth of a quarter of a mile, darkness, to our eyes, reigns supreme. But delicate photographic plates are affected well beyond half a mile by the chemical rays at the farthest end of the spectrum. Even on the blackest midnight on land there are ultra-violet rays playing everywhere, but a mile down the darkness is absolute; the temperature has lowered many degrees toward the freezing point and now on every inch of surface there is a terrible pressure of over a ton. Down and down we sink, our feet touching bottom, in some places, over six miles below the warm, sunlit surface. Even here weird worms, fish, crabs—uncouth and unearthly, live out their lives in the midst of eternal silence, blackness and quiet—feeding on the refrigerated remains of animals, which fall from unimagined regions overhead.

Although the sun is wholly blotted out, yet as we get light from coal fires on the darkest night, so in the depths we distinguish dim lights here and there, and for mile after mile the great watery spaces are faintly illumined with the yellowish-green glow from countless millions of living candles on the skin and scales and fins of wandering fish, worms and shrimps.

With the passing of the warmer light rays, plant life ceases, so below this point all the living creatures are carnivorous, and beneath a certain depth they become subject to a death so terrible as to seem appropriate to these regions. If from injury or other reason their tissues develop gases, they begin to fall upward. Once beyond the pressure to which they are used, they roll and twist up and up—gravitation for the moment helpless, and finally expire from the heat and light and lowered pressure, and float at the surface until devoured by fish or bird, or captured by some lucky scientist.

I was led to believe that all the deep sea creatures would come up disfigured, with internal organs forced out, eyes displaced, scales gone. But for some reason good fortune was with us and again and again deep sea fish and other organisms lived from two minutes to as many hours,—and swam and breathed and sent forth barrages from their luminescent batteries—the strength of which sometimes lighted up the whole dark-room where I studied them.

Now and then there was enacted some little dramatic incident before my eyes which revealed the ways of life in this underworld. One of the best known camouflages is the trick played by a squid when threatened. He shoots out a dense cloud of sepia ink—a most efficient smoke screen which, in the sunlit, surface waters, wholly blinds any assailant, and in the ensuing confusion the squid darts off backward to safety. No better plan could be imagined in sunlight, but how futile such a habit would be in the stygian darkness six hundred fathoms down. Not far from New York City I took from that depth a scarlet prawn two inches long (Plate VII). As usual I put it in a large jar of water and rushed with it into the dark-room. When my eyes became accustomed to the darkness I watched carefully and long, but not a flicker or glimmer could I perceive. Just as I was about to give it up I saw a dull glow from what I took to be some one-celled organism, perhaps a dying *Noctiluca.* To my astonishment it increased in size, and, bringing near the illumined face of my watch, I saw the source of the fiery flow was the prawn itself. The light now took the form of a liquid pouring out into the water, and soon the entire contents of the aquarium was aglow, while, swimming about in it, the prawn could be seen as a black, inchoate mass. Suddenly the significance of this occurred to me—this red crustacean was playing the same trick as the squid, but adapted to the darkness of six hundred fathoms. The squid had its cloud of smoke by day, the prawn its pillar of fire by night.

In any consideration of the sea from surface to bottom we must not omit color, and it is possible to distinguish several very generalized zones, often ill-defined or overlapping. In the sunlit strata we have the ultramarine and the transparent creatures, such as flyingfish and the shelless *Glaucus,* the strange *Leptocephalus* eel-lets and the infant lobster ghosts. Then there comes the silver zone where live many fishes gleaming like molten tinsel. Next the area of pink colored life, and last of all the beings clad in scarlet and in black. Red, of course, can be a color only in light, but as a matter of mere pigmental economy we find a host of scarlet animals living alongside the jet black ones. Now and then there comes up a stray fish or worm or sea-cucumber as pallid as a sunless plant.

Fig. A. SQUID THROWING OUT A DEFENSIVE
SMOKE SCREEN OF SEPIA INK

(One-half natural size)

Fig. B. DEEP SEA PRAWN EMITTING A LUMI-
NESCENT DEFENSIVE CLOUD

(Twice natural size)

PLATE VII

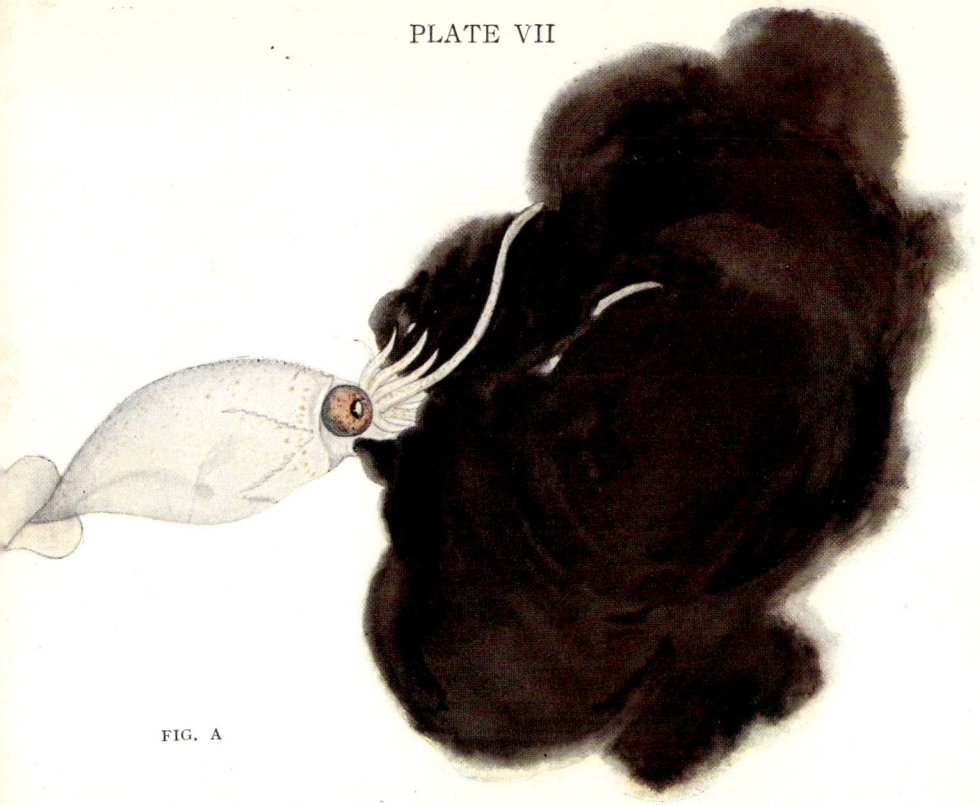

FIG. A

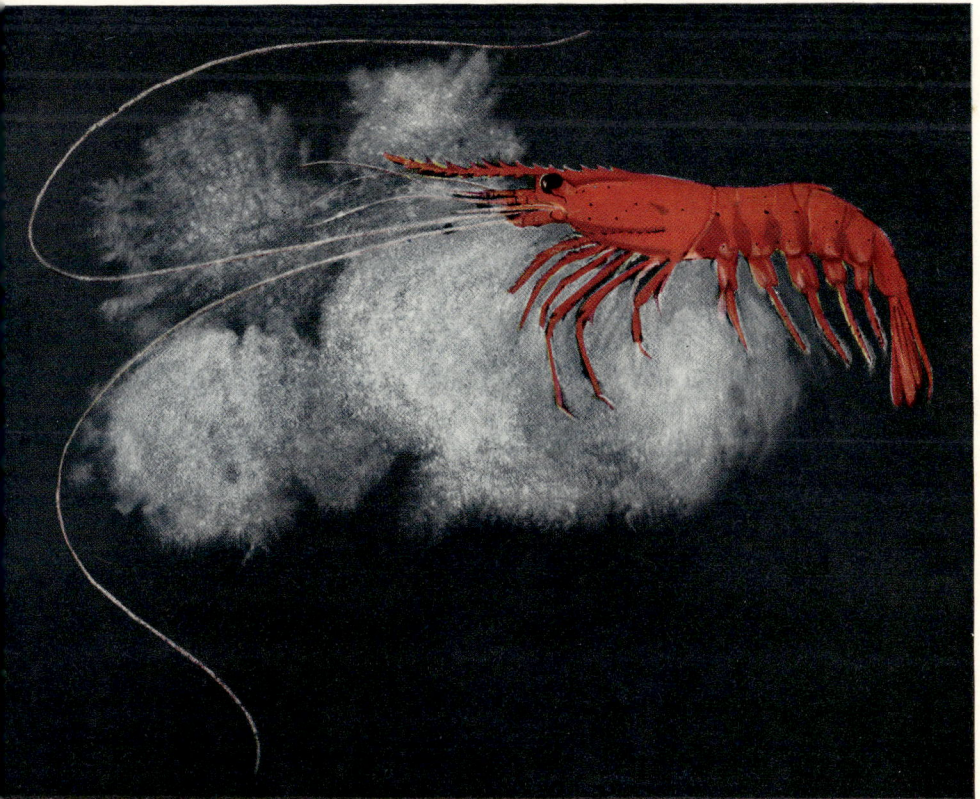

FIG. B

Always the lamps of the undersea host held the chief interest. We can understand a fish like the coppery lantern-bearer with many lights, and normal eyes to take advantage of the illumination. *Argyropelecus* (Frontispiece, Plate VIII) however, is the first of many deep-sea puzzles because, while the lower sides are lined with large luminous organs, the light from which is thrown downward rather than sideways, yet the eyes, which are very large and bulging, are directed straight upward. Why this fish should be denied the ability to enjoy its own pyrotechnics is not apparent. If the downward sheet of light acts as a lure to attract its prey there still seems considerable need for anatomical alteration, for the mouth, like the eyes, is turned almost straight upward. Twice I secured living specimens and three times I was able to distinguish the illumination.

Often in the same net with the silver hatchet was a still stranger looking fish, *Sternoptyx* (Frontispiece, Plate VIII). It is impossible to describe except that the shape seemed all wrong. When I saw the first one I was certain that this was one of those distortions of which I had read, due to lessened pressure, but I soon realized that the fish must live happily with a body outline like nothing else in the world. The head was fairly normal, but the anterior half of the body was dragged downward twice as far as it should be. Then just when I was willing to accept this outline and follow it along to the tail, I found that the posterior half of the body was again all wrong—

being reduced to half its height and jammed up against the dorsal fin. What at first glance was the lower portion was seen to be only a thin layer of quite transparent tissue, through which visibly extended various bones and fin rays which in any more correctly made fish are always decently concealed within the body. I felt like applying Buffon's opinion of sloths, that if it had one more defect, it would cease to exist.

When we find ourselves in an egocentric mood such as this, we have but to think what comment *Sternoptyx* would make on our own figure were we to drift down past him in the darkness of his deep home. He had not nearly as many light organs as his cousin *Argyropelecus,* and they were scattered in patches of twos and threes here and there over his much-angled body. Judging by his color he was a sharer of two zones, a coat of black pigment being overlaid with a tissue of silver. Only once was I fortunate enough to see a live one, which swam feebly in circles for a few minutes.

Another confusing condition of affairs came to light (in every sense of the word), when I found a brightly illumined blind fish. The lights may have persisted from the time when its eyes were better developed, but a more probable explanation is that the rays act as a lure for small edible creatures, and the fish, through sensations other than sight, is able to detect their presence and to seize them. Until we actually know the cause however, we can only speculate, and allow it to bring such absurd similes to mind as a blind Diogenes stumbling along with a lighted lantern in his hand.

From four hundred and fifty fathoms, or about half a mile, I took several fish at Station Seventy-four, of the euphonious name of *Bathypteröis* (Fig. 47). For these we can find a more reasonable simile—that of a blind man walking down the street and tapping with his cane ahead of him as he goes. They were good-sized fish, six inches to a foot in length, black as usual, and although the eyes were present, they were exceedingly degenerate and apparently useless. The pectoral fins were compensation, being split up into numerous, elongated rays, the lower ones of which reached almost as far back as the tail. When spread out sideways these formed a great sensory portière, while the upper one on each side was still longer and divided at the tip into two feelers, so that these controlled a still wider field of touch. In lieu of eyes, these many fingers enabled the fish to obtain food, to avoid danger and to find its mate— and when this is said and done the destiny of a fish is accomplished.

Let us turn from fishes for a moment and go out on the deck of the *Arcturus* in answer to a shout concerning an incoming net. The great silk cone rose dripping from the waves, and at the apex I could see a sagging mass of pale salmon jelly. This I carefully decanted into a white enamel pan and carried into the laboratory. The mass was icy cold, and no wonder, for it had been strained from waters three-fourths of a mile beneath the vessel.

It was semi-solid and looked as if made up of a thousand bits of parti-colored glass and jewels,— the living loot, motionless and inchoate, of some abyssmal Aladdin's cave.

I poured water upon it, and with the dilution came disintegration of the plankton. As the tens of thousands of atoms of jewelled jelly fell apart, each assumed the shape and character of a complete individual, and at once began to kick or breathe or swim or throb after its kind. Visible blood started to circulate, hearts were distinctly seen to pulsate in the depths of glassy bodies, enemies leaped at one another's throats (or whatever they possessed in lieu of such a region), and on the instant of watery liberty, feathery-footed males danced and whirled in courtship ecstasy about their less ornamented mates. These are no meaningless, flowery phrases, for within a minute after dilution, the jelly mass under my binoculars exhibited every emotion known to invertebrates.

The general color of this concentrated animal life was a rich salmon, picked out with spots of dark brown, black, maroon, purple, and with scarlet so deep and vivid that it instantly attracted and hypnotically held the eye. Near one spot of this violently insistent color an acrobat drew my attention. Although he was floating freely, his actions seemed curiously limited. He might have been clinging to an invisible bar and doing gymnastics on it. I spooned him out and his restraint became understandable. Again I call on Diogenes for comparison, for it was as if that cynic philosopher had been doing hand-stands and somersaults in his cask. Here before me was an amphipod in a barrel, —a transparent barrel to be sure, but one which had been well hooped and staved by some cooper of the underseas. As for the crustacean itself, it was one of a group which Latreille, a century and a quarter ago had named *Phronima*. A second individual in another barrel was wrapped in a veil which the lens resolved into a host of pink, infant *Phronimas*. So this was no casual or accidental association, and if I could have watched one of these youngsters, I would have seen it in the course of time seize in its turn upon a barrel as it floated past. This barrel, by the way, is the shell or test of an ascidian, *Doliolum* by name, a creature who, in common with its relatives the salpæ has slipped down the evolution ladder a few rungs, after actually coming within sight of the vertebrate goal (p. 380). *Phronima* recked nothing of this and proceeded literally to eat *Doliolum* out of house and home, and to climb in the back door. Not only did she thus acquire a glass house and a nursery to order, but a motor boat as well. Clinging by her largest pair of claws, she stretched her body far out behind, and by a frantic fanning of the water was able to get up astonishing speed, at the same time forcing the water in at the front door, bringing with it oxygen and food for herself and her brood.

The head of *Phronima* was like nothing but the head of another of its own kind. Its overbalanced appearance reminded me faintly of a termite, but its eyes were well worthy of the cranium in which

they were placed. On the summit of the head were
myriads of tiny, bead-like facets, each on the summit of a long thready nerve extending down to the
side of the head where were two more normal appearing eyes. It seems that *Phronima* is especially
blessed with eyesight, for it is believed that the first
mentioned structure functions in the pale, dimness
of the general environment, while the lateral eyes
are better fitted for focussing on brilliantly lighted
objects. If we can imagine one of the voracious
Astronesthes coming along in the full blaze of its
hundreds of portholes, *Phronima* at once brings its
side eyes into play, puts its tail hard aport, and
goes off in its barrel at high speed.

Every half inch shift of the plankton pan
brought a new world into view, or rather a new
cosmos, for along came a planet rolling slowly
across space—a wine-colored sphere of a jelly-fish
—as heavy as a shadow, as dense as water, as beautiful as could be. It seemed very far from the animal kingdom, with its radial symmetry of a flower,
for which it could blame, or boast some long-
stemmed ancestor of ages past. Sagittæ, those
swift arrows of sea-worms, showed the trace of their
deep home by their pink hue—an approach to the
scarlet and the black zones of water. They hung
motionless in the quiver of their own body, or shot
with half-sheathed jaws swiftly through the mass
of plankton. They are the falcons of this plankton world, and in the stomach of one I found a lantern-fish, a *Vinciguerria,* as perfect as if it still
lived.

A moonstone, cut in the form of a smooth, exquisite oval caught my eye, and I found it to be the test of a siphonophore. Under the lens it transcended the beauty of any inorganic jewel, for it throbbed with life and revealed most intricate structure. Its substance was as evanescent as a mass of intersecting shadows caught prisoner for a time in the meshes of a few drops of salt water,—the curving muscle bands, the many infinitely minute rods, cunningly braced, the inward dipping mouth —all were perfect, and the play of color over the surface surpassed the iridescence of soap-bubbles.

Stirring up a mass of dull grey plankton, again there came the shock of sheer color—like a blow to the body, or a crashing chord to the ear. I know of no other sensation which quite equals the effect on the eye—or the brain behind the eye—as that of a great, glowing, living, rich-scarlet-red shrimp, cold as ice, just raised through a half mile of water. No flower I have ever seen in any setting could vie with it for a moment. It is worth recalling that for countless ages this shrimp and its ancestors had been merely the blackest of beings in a jet-black world, and only for the past few minutes had its blazing color existed. This may partly explain its exciting quality, like the unused rods and cones in our own retina, when we stand on our heads and look out at the world.

When an unexpected roll of the *Arcturus* washed the main mass of the cold plankton to one side of the pan, there remained on the bottom a thick deposit of a myriad, fine dots, of all colors and sizes. This was, if possible, a more beautiful and astonishing world of life even than the larger creatures drifting overhead. Here were hundreds of shells of the one-celled globigerina (Fig. 11), all with their minute occupants, amœbic blobs of protoplasm. Other shells were rounded, or elongate or heliced, and with them were mingled a fewer number of real snails, some bivalves and some turreted. Here and there, from a nautilus-like shell, an animal something began to protrude. Out of a spiralled mass of unrecognizable tissue there slowly emerged two eyes, a long proboscis, and, in the wake of several other organs, a pair of wings. As I watched, the wings began to flap, slowly at first, then with more force and regularity, and the snail, shell and all, rose slowly from the globigerina and went flitting off through the water like a rather unskillful bat. Here we have the secret of the molluscan life half-way between surface and ocean floor, and again the deeps show what they can do in the way of miracles—flying snails!

Now came two creatures, signalizing the antitheses of life in these regions. My flying snail and a thousand tiny copepods and sagittæ were suddenly shouldered aside under my eyes by a moving rainbow—a jellyfish without a shadow, and it in turn was pushed out of sight by a very small but very terrible octopus, black as night, with ivory white jaws and blood-red eyes. This came along, half swimming, half sidling, its eight cupped arms all joined together by an ebony web. In those icy, black depths, to be a small fish and to come within reach of such sinister arms, to be enfolded by the living umbrella, and then drawn slowly, irrevocably toward the wide-open, gleaming beak, watched always by those cruel, lidless eyes, so frightfully like those of human beings, seemed, to my imagination, a much more awful fate, than could ever befall, in our darkest night, any creature breathing air.

So much for the hastiest glance at a plankton haul at Station Seventy-four. The following day, at the same Station, we made a memorable haul on the bottom, six hundred odd fathoms down, and took a whole tubful of bottom fish, brotulids and macrurids, together with a gorgeous lot of invertebrates.

Of the marvellous hauls of crustacea, or crabs, shrimps and prawns, which we made at this Station, it is most difficult to write, for the majority are new or exceedingly rare forms, and almost none have any common names. Lee Boone who, in a masterful manner, is studying them from the technical point of view, has, at my request, written a few paragraphs of the general impressions of this unusual collection which I am glad to reproduce here.

One of the most interesting of such hauls is illustrated in Figure 55, which shows some of the loot from Station Seventy-four, Otter Trawl No. 3, Depth 624 fathoms. In this were "opalescent *Gigantocypris,* as astounding a surprise as if one met an ant a yard long; scores of scarlet armored *Heterocarpus* with swordlike rostrum half as long as the body, and slender, sensitive, wavering antennæ three or four times the animal's length; transparent amphipods who, paradoxically, found their strength in fragility; an elegant amethystine crab whose lilac-grey eyes were set at the end of long flexible stalks that enabled the eye to be swung in an arc with a radius of half the body length, above or before the creature, periscope fashion; a weird, eyeless, claw-footed *Willemœsia* that has curiously retained the characteristics of his long vanished kin of the ancient Triassic seas; a new species of *Uroptychus,* with exquisitely sculptered body of pearly safrano pink and queer long claws that dredge the ocean floor in search of food; hundreds of swift *Benthesicymus,* vivid crimson slashed with spectrum blue; an apparently headless, multispined orange amphipod—a miniature impersonation of some dread prehistoric monster; countless scarlet *Nematocarcinus* whose fantastically long legs were so shadow-thin, one half-expected them to ballet Stevenson's nursery song; a small, globular, porcupinish *Eryonicus,* who had solved the problem of living in the great depths, where the pressure is more than two thousand pounds to the square inch, by evolving a spherical form and a flexible leathery coat in place of the usual rigid armor worn by his kind; a delicate ivory-hued *Paribaccus* tinted with sea-foam green, seemingly as fragile as a bas-relief on a bubble, yet encompassing with its shadowy form the multiplicity of segments and the complex nervous, circulatory, respiratory and other systems that coordinate to conduct the business of living; a group of comical hermit crabs—absurdly grotesque clowns who had cunningly hidden their weak, misshapen bodies under the deceptive, flowerlike, death-dealing tenacles of rosy sea anemones, while everywhere, jewelling the nets like fragments of girascole, were the little Ostracods.

"Rendered conspicuous in this colorful throng by the neutral tones of its monk-grey garb was a small globular crab which seemed at first glance as immutable, as lifeless, as a bit of Archean rock from the ocean floor, but which, upon closer inspection proved to be one of the most remarkable crustaceans captured by the expedition. It is a new member of the trible *Dromiacea,* that curious group of primitive, sponge-carrying crabs of the West Indies. Like its shallow water relatives this species also clothes itself in foreign substances, but instead of sponges it uses minute animals, globigerina, sponge spicules and sand grains. These are held in place by remarkable tree-like hairs which cover the entire crab. It has evidently long been an inhabitant of the abyss, for the eyes are small and degenerate and the antennæ are exceedingly long and tactile. And finally and most unexpected, situated at the base of these antennæ and opening just in front of the mouth cavity are the ducts from paired luminous organs. When released by the opening of the magical circular door which is formed by the first joint of the antennæ (a segment lost in most crabs) the luminescent substance glows like a tiny lantern, and may well serve to attract a host of small creatures who are promptly devoured.

"Finally we espied the most supremely interesting crustacean of the entire expedition, who with the unobtrusive modesty of the truly great, made no flaunting claim for attention, a shy, small creature which might well have been overlooked as one of the myriad amphipods surrounding him. Yet a subtle prescience of immemorial mysteries proclaimed him to be a Missing Link. This is one of the most bromadic of phrases, and is misused ninety-nine times out of a hundred, but in this particular case it is as well deserved as it can be when applied to any creature still living on the earth today. We looked, afraid to look, mutely questioning the unbelievable evidence. It was more inscrutable than the Sphinx as it was unquestionably æons older in its characters. With fear lest our treasure vanish in the dispassionate light of scientific fact, we examined this exquisitely delineated ivory figurine.

"Could it be an aberrant amphipod? Yes, suggested the side plates and some of the appendages, but no, declared the stalked eyes and macruran-like carapace. Could it be a macruran? But the carapace is composed of seven *articulated* plates, and the first abdominal segment has no counterpart in macruran morphology. Vaguely, it calls to mind the anterior abdominal segments of a *Lithodid* crab. Is it a crab? Still more impossible.

"Segment by segment we analyze the mystery, and tabulating the data, find that we have a primitive macruran crustacean, so aberrant from all known forms that a new family must be established

FIG. 53.—A LIVING SEA DEVIL, *Melanocetus* sp., PHOTOGRAPHED WITHIN A MINUTE OF CAPTURE. FROM A DEPTH OF 600 FATHOMS.

FIG. 54.—A BROTULID FISH STILL LIVING AFTER HAVING BEEN BROUGHT UP FROM A HALF MILE DEPTH.

for it. And so the Missing Link is christened, *Probeebei mirabilis,* for we were convinced that Father Neptune had sent a special greeting by this messenger to his brother-in-fins William Beebe, the Director of the *Arcturus* Expedition."

Among the fishes in this haul were eight little sharks hardly as long as one's hand. They were slender, with dull glowing eyes of emerald green and with stout spines in front of the dorsal fins. Delicate white markings were difficult to account for except as a heritage from some ancestor who lived where there was sufficient light to permit a sharklet to see whether he was white or black. I found they had been feeding on small scarlet prawns of a species which we had not been able to capture. In the same net were several giant shrimps so armed and armored that in a battle between them and one of the sharks, the latter would undoubtedly have been worsted.

In the midst of all this richesse, a compound thrill was vouchsafed, when in the midst of a mass of grey sponge and a scarlet haze of shrimp antennæ, there shone out a rainbow glint, and I uncovered a half dozen large snail shells of solid mother-of-pearl luster. They were quite dead and their food was mud and globigerina ooze, and yet in the cold blackness almost three-fourths of a mile below the surface, the living snails had been clad in a gay livery of orange, green, black and white.

If I should consider the deep sea and its inhabitants solely from the point of view of technical science, I could never use such words as terrible, strange, beautiful or ugly. Because, in the essence of things, a pressure of a ton on each square inch is only a normal shift in physical conditions,—a fish which is chiefly mouth is merely a specialized adaptation to its particular environment, as is the smaller organ of our brook trout. But in this chapter we may let emotional appreciation go hand in hand with truth, and science will take no harm. To consider variety of mouths only, Figures 51 and 52 show what came up in a single haul,—the great cavernous maw of a pallid-white pelicanfish, and the unbelievably thin and curved, wire-like jaws of a silvery thread eel.

Turn please, to the little sea-devil in Figure 1, and be honestly astonished enough to exclaim something more than *Diabolidium arcturi!* although, come to think of it, that does have the advantage of sounding like a hearty exclamatory oath.

As I have said before, with the passing of red light and plant life we descend into the zone of carnivores, where every living thing is compelled to feed on other animals, living or dead. I am no vegetarian, but when I see a mighty ox or elephant or behemoth himself in full action I do not belittle the brawn- and muscle-making possibilities of a plant diet. But the gentleness of countenance of a cud-chewer, the soft, mild eyes of kine are proverbial, and when I realized that tooth and claw reigned supreme in the dark under-water world I wondered whether this diet would affect the mien of the natives.

Without further preamble, we can safely assert

it does, although we are only on the threshold of intimate knowledge of the life histories of these sea creatures. Circumstantial evidence, however, is often conclusive enough proof, as Thoreau said when he discovered a trout in the milk, and when we bring up a fish which has swallowed another five and a half times its own length, we realize that we are far indeed from the seaweed nibbling zone.

All deep sea life has either slid slowly down the continental slopes or year by year become waterlogged to deeper and deeper zones from the surface of the sea. Hence we often find relatives of the abyssmal forms quite near home. The angler is a common fish which buries itself in the mud, with a long, fleshy-tipped tentacle lure dangling freely in the water. At the approach of prey, almost the entire fish opens into an enormous mouth and engulfs the unwary victim. The capacity and voracity of its deep sea relatives are adumbrated in this shallower water fisher, for seven wild ducks have been found in the stomach of one of these fish.

In the illustrations of this volume I have included seven deep sea anglers or sea-devils (Figures 1, 48, 49, 50, 58, 59, 60), to show the variety of these remarkable fish. Most of these are unnamed species but to the first (Figure 1) I have already[1] given the name of Little Devil of the Arcturus, *Diabolidium arcturi* and the rest will already have been christened by the time this appears in print. Figures 53 and 54 are photographs

[1] *N. Y. Zoological Society Bulletin,* XXIX, No. 2.

of a sea-devil and a brotulid from even greater depths, taken while the fish were alive, breathing and swimming. These sea-devils differ from the shallow water anglers in being rounded rather than flattened and this shows that they are not bottom livers but mid-water floaters. In fact, some of them, such as *Diabolidium,* could not very well rest on anything hard without damaging some of their delicate structures. Most of those we captured were a hundred or two hundred fathoms at least from the sea bottom. In one haul at Station Seventy-four we took seven individuals of six species. Of the seven illustrated three were alive when they came to the surface, and two showed distinctly illumination of the bulb-like tips of the tentacles. In *Diabolidium* not only the tip of the tentacle, but all the larger teeth were dimly outlined with luminescence—apparently a mucus like that given off from the numerous pores of the skin. I tried to estimate roughly the relative proportions of the mouth and the rest of the body and in two species found it quite four-fifths of the entire animal.

And so, had we space to go on, we might show that a generous majority of the deep sea fish are little more than living eating-machines, with every function subordinated to that of capture with the appalling rows of teeth, engulfment in the cavernous mouth, and finally reception and digestion in a stomach which is beyond belief elastic and distensible.

I shall conclude my notes on these deep sea fish with the account of a discovery published [1] about a

[1] C. Tate Regan, Proc. Royal Soc., London, XCVII, No. B684, page 386.

week before we sailed on the *Arcturus*. This is, to my mind, the most remarkable and unexpected result of deep sea investigation, and reveals a condition existing in backboned animals which elsewhere is found only among such lowly organisms as certain crustaceans and worms. Three genera of these abyssmal sea-devils have been found in which a diminutive adult male fish is actually growing from the side or head or behind the gill of the female—a parasite in every sense. So complete is the association that the male derives all nourishment through the blood supply of the female, and hence has lost teeth and the luminescent lure, while eyesight and the alimentary canal have degenerated to the vanishing point.

Stimulated by this news I scrutinized every specimen of these little sea-devils, but was not fortunate enough to find such an association. When, from this astounding example, we realize the possibilities of deep sea life still unknown to us, every haul of the dredge should be welcomed by an expectant enthusiasm equalled in other fields only by the possible hope of communication with our sister planets.

CHAPTER XV

FISHING IN THE HUDSON'S ANCIENT GORGE

AT four o'clock in the morning of July 25th I was on the bridge of the *Arcturus* when the Captain signalled for slow speed. For an hour we barely pushed through the water, while two sextants were levelled at our namesake which glowed brightly in the heavens. Finally a pencil made a tiny dot on the chart, Full Stop clanged in the engine room and we floated quietly over our objective—the sunken gorge of the Hudson River. There was just a hint of dawn in the star-flecked east as I went to my cabin for an hour's sleep.

There are mirages and illusions of the senses and there are those of the mind, and in the full light of day I found myself laboring under both. Our last mainland sighted was the old, pirate-famed harbor of Porto Bello. By solar and sidereal observations we had been close to Chesapeake to make connections with the *Warrior,* and dredged there in fifteen fathoms with no hint of land in view. Now we were one hundred miles from the New York City Hall, according to the word of the Captain, and in six hundred fathoms of water, according to the sounding wire. I found it quite impossible to realize that my city was only an hour away by plane and a day by steaming. The sea stretched unbroken to the horizon just as it had done week after week, and month after month in the Atlantic and in the Pacific, and our senses and our minds insisted that we were still thousands of miles from anywhere.

A recently conceived plan only added to this conviction of distance. Our homeward-bound pennant with its one hundred and eighty feet of length, for the hundred and eighty days we should have been away, was furled, ready to be broken out, but no thought of packing had entered our minds. We were all still in woollen shirts, khaki shorts and sneakers which had been our entire garb for half a year. The odious stiff collars and shirts, the silly colored strings of neckties, the funereal dinner jackets, together with all the other uncomfortable and unlovely portions of civilized attire were still packed away, snuggled among moth balls in the hold. My plan was that our last station—Number One Hundred and Thirteen—should be here in the depths of this royal gorge of the Hudson River, within reach of what was once by far the greatest waterfall in the world, and today a scant hundred miles from our city of New York.

I was about to grope beneath half a mile of water for vague hints of whatever life the fingers of my dangling nets might bring up, and so it seems not unreasonable to look back through past ages to the time when this gorge roared with the thundering stream of the Hudson, and to attempt

with pitifully feeble gropings of the imagination
to repicture some of that distance scene. My data
is all based, of course, on geological volumes, and
appears to be accepted by most reliable students of
physical geography.

As nearly as we can judge, the period of the
early Pleistocene was something like a million
years ago, and at this time the northeastern coast
of the United States was elevated a mile or more
above its present level. This made of Manhattan
Island an elongated line of rugged hills about one
hundred miles inland, while the great Hudson
drained not only its own valley, but most of the
great lakes to the north. This mighty flood rushed
southward through the Palisades, past Manhattan
and on out toward the Atlantic, augmented by
the tributaries of the Connecticut, the Housatonic,
the Passaic and the Hackensack Rivers.

So low has the coastal region sunk since that
time that today the Hudson, as far up as Albany,
is little more than an elongated fjord, the effect of
the ocean's tides being felt throughout this entire
distance of a hundred and fifty miles. Even the
Palisades were much more imposing in olden
times, for the glaciers had not then filled the
Catskill bed of the river with the hundreds of
feet of rocks and gravel which now choke it. If
we could have then floated down the Hudson the
Palisades would have towered four times as high
above us.

In those days the compound river rushed
through the channel which on clear days can now

be seen from an airplane as a dark streak beyond
Sandy Hook. For a distance of forty-five miles
beyond what is now dry land, the Hudson flowed
rapidly but evenly through a fairly deep bed, between the level banks of the wide, sloping coastal
plain. Then, without warning, its waters plunged
into the maw of a canyon mightier than man has
ever seen. At the head it was less than a mile wide
and rapidly reached a depth of sixteen hundred
feet. Today our sounding line touches bottom
four hundred feet down on the surface of the ancient plain, while a few hundred yards away the
plummet sinks into the gorge to a depth of twenty-
eight hundred feet. Four miles farther down the
canyon, where the land of the ancient coast is now
a thousand feet under water, to reach the bottom
of the gorge requires forty-eight hundred feet or
almost a mile of wire. Here the entire volume of
the Hudson, plus the lake water and the tributary rivers dropped almost sheer over a precipice
of more than eighteen hundred feet—more than a
quarter of a mile. The only thing on the earth
today to compare with this is Kaieteur Falls in
British Guiana. This has a maximum drop of
over eight hundred feet, the highest waterfall,
worthy of the name, in the world today. To the
chosen few who have seen this, the mind is able
dimly to repicture the incomparable gorge of the
Hudson as it once was. The thrill which came up
over the vibrating piano wire when we touched the
very bottom, brought to the imagination what the
most marvellous piece of music conveys to the

ear. It was a lost chord vibrant with all the wonder of past ages, before man or his kindred had begun to evolve.

During the successive glacial ages when time after time the enormous masses of ice advanced and retreated, the coast slowly sank and, before the end of the Pleistocene Age, presented a contour much like that of today. During all this period the wild life of Manhattan and the adjacent country was diversified and wholly different from that of historical times. As the climate alternated from Arctic to semi-tropical, successive faunas replaced one another. At Long Branch there lived during widely separated times, such unlike creatures as walruses and giant ground sloths. Mastodons were abundant even on Manhattan, while not many miles from the Hudson were wild horses, tapirs, peccaries, reindeer, muskoxen, bison and giant beavers. Most of these animals lived long before the first evidences of mankind, and the great submarine canyon was never seen by any eye of man or his immediate forebears.

And now instead of thinking back through time forever lost to us, I was about to reach down through space equally forbidden to living man—into a region comparable to the ether beyond the neighborhood of comfortable planets and world sanctuaries, a region eternally cold, with ultimate silences, and darkness and pressure beyond all human imagination.

When our soundings revealed the fact that we were actually floating over the deepest part of the gorge, and had reached the point nearest New York City where we might expect to find the strange creatures of the abyssmal depths, I gave orders to put out the string of nets which had yielded the best results during the past months. First there was paid out the otter trawl, a great bag of netting forty feet in length, with its gaping mouth held wide open by the oblique pull of two iron-bound boards. Then, at intervals of fifty fathoms, meter nets were lowered, each twenty feet long, made of the finest, most costly silk, with a mouth composed of a brass ring a yard in diameter. Five of these nets were attached to the steel cable by guide ropes, and they trailed straight out behind at the various depths as the ship steamed at slowest speed through the water. For three hours they were pulled gently along at 500, 450, 400, 350 and 300 fathoms depth, blindly, uncontrollably but usually successfully engulfing the weird beings which happened to float along in their path.

Although, as I have said, the expanse of open ocean conveyed no hint of the actual nearness of land and human beings, yet hardly had the last net disappeared beneath the surface when ships appeared on the horizon. A square rigger drifted slowly along with slack canvas, while at her heels followed casually but watchfully a low subchaser. A line of smoke in another direction marked a dainty white revenue cutter which came tearing full speed toward us. We chuckled as we thought what a suspicious-looking craft we must be—all

begrimed with the outboard trawling, six months
of weed on our keel, and rolling in the swells for
no apparent reason except for an inexplicable steel
cable leading obliquely down into the blue depths.
We rather looked forward to the excitement of
keeping up our mysterious character until we were
boarded by this bootlegger policeman. We even
anticipated offering the officer a cocktail, thereby
breaking no law of which we were aware—being
one hundred miles out at sea and having gauged
our Panama supply exactly up to the last moment
before landing in New York. But the cutter's
captain knew what he was about and had evidently
been expecting us, for as he encircled us he dipped
his ensign and saluted us with the usual three
blasts. The unexpected compliment thrilled us
and we answered with the deepest basso profundo
roars of which our whistle was capable.

During the succeeding four days and nights
which we spent drifting over the gorge we had not
a moment's idleness from lack of specimens.
Throughout the day we kept up constant trawling or dredging, and at night trawled with small
surface nets or harpooned and netted fish and
other creatures from the pulpit and gangway.
Even before the stormy petrels discovered us, we
were a source of food supply; the sharks came and
circled us eagerly—not in hopes of any human who
might by chance fall overboard—I had exploded
this myth pretty thoroughly in my intimate association with them during the last six months—but
on the lookout for garbage.

The sailors borrowed some of my shark hooks and chains and in quick succession caught three over the stern. All were *Carcharhinus obscurus* —the dusky ground shark which seems to be almost unknown near New York, although common to the north at Wood's Hole. The most noticable character of these creatures was the pale color of the fins. The pectorals were greyish-white for half their length and when swimming in the sea they appeared milk-white. These sharks arrived singly and converged toward the bow and then drifted sternward. Perceiving the slowly dragging bait, they leisurely swam toward and engulfed it, with, however, none of the story-book legend of having to turn over on their backs before seizing their prey. A male shark measured over seven feet in total length and weighed one hundred and twenty pounds, after we had all estimated his weight at about three hundred!

At Porto Bello we had purchased two small puppies of doubtful, or rather of quite certain absence of, pedigree. They were most amusing little fellows and were thoroughly spoiled by everyone on board. Both, unfortunately, developed signs of mange and much to their disgust we treated them thoroughly with the old reliable Glover's. They had grown and thrived apace, but now the smaller of the two pups, Blanco Ugly as we called him, by accident or intention (the Spanish-American temperament being so uncertain) fell overboard and drowned before anyone could see or save him. The first we knew of the tragedy was the sight of his little body drifting alongside the almost motionless vessel. Immediately a great shark rose beneath him, engulfed him with a single effort and sank from view. Quickly, however, as this had taken place, the shark reappeared and relinquished the puppy intact,—Glover's mange cure apparently not appealing to the palate of this scavenger of the sea.

The previous day we had received a generally broadcasted wireless, warning ships to be on their guard against a derelict—a schooner which had been run down in our vicinity but not sunk. This was brought to mind when a hatch drifted past, then a chair and pieces of masts and rigging. Once a huge squared beam was sighted which at first we took for an upturned ship's boat. I put over a small motor boat and the two men who went out to the floating object reported that the beam had been adrift for a long time, as it was covered with barnacles and weed. A host of fish swimming beneath it tempted me to use the last few sticks of dynamite which we had left. A number of fish were killed by the explosion but all sank at once or were taken by sharks before we could secure them.

Whenever the vessel was moving we trolled with spoons and artificial squids for stray tunnies and mackerel. Large swordfish came several times to the shimmering bait, and one even tasted it, but the slightly irritated nod of his head parted the stout cod line as if it had been cobweb. I record all these casual occurrences to indicate the many ways in which it is possible to capture specimens at sea in addition to the usual nets and dredges.

Before we return to examine the contents of our deep sunken nets and trawls let us see what the surface has to offer us as we float where, long ago, great primitive eagles soared and looked down on ancient landscapes. In relation to those days this present year is more nearly 1,001,926.

The larger surface life was abundant and schools of tunnies passed now and then, looking from the deck like flocks of violet torpedoes, while dolphins came and inspected us, and went on their way rocketing. We watched one which never failed to leap high and fall back flat on his back with a resounding slap. If it was play he was a confirmed humorist, if unromantically merely to dislodge barnacles or parasites from the skin of his back, he must assuredly have been successful. The most impressive visitors were schools of small whales or blackfish, which rolled in a dignified, elephantine manner through the waves and with huge sighs sent up spouts of mist.

Next to the general oceanographic machinery of nets and dredges the apparatus most constantly in use was the metal front porch or pulpit which we let down over the bow close to the water. This was seldom vacant during the day and when aquatic loot was abundant two of the staff sometimes worked in it at the same time, with long-handled net and pail. The objects thus captured floating over the sunken Hudson gorge varied from scientifically rare to beautiful to merely comic. Christo-

pher Columbus hailed birds and floating grass as
indicative of land,—so the comic elements in our
pulpit hauls adumbrated human proximity.   Here
is a catalogue of these items taken on the first day,
showing a pronounced lacteal dominance:

Rubber nipple from a baby's bottle	1
Cardboard milk-bottle tops	4
Empty milk of magnesia bottle	1
Cans	2
Leg of rubber doll	1
Piece of bath tub	1
*Empty Gordon bottle	1
Large wooden spigot	1

*Possibly autochthonic to the *Arcturus*

We were well inshore, away from any strong
influence of the Gulf Stream, in an eddy-like backwater with little current, so that we found creatures which had drifted out of the main Gulf
Stream, as well as others which hailed from the
shore.   Although there was no strong offshore
breeze, yet an astonishingly large number of insects had found their way these hundred miles
from land, and we captured thirty altogether, including moths, grasshoppers, beetles, and dragonflies. Some were struggling their last in the water,
others flew wearily aboard the *Arcturus*.

Scattered bits of sargassum weed floated here
and there—sad little plants of the sea, for all were
doomed. Better for them if they had clung to the
northward flowing stream, within a few days to

FIG. 56.—DUSKY SHARK CAUGHT ONE HUNDRED MILES OFF NEW YORK CITY.

FIG. 57.—A DEEP SEA SHARK, EIGHT INCHES LONG, FROM OVER A HALF MILE DEPTH
AT STATION 74.

FIG. 58.—A SEA DEVIL WITH LUMINESCENT TEETH.

FIGS. 59, 60.—TWO SPECIES OF SEA DEVILS FROM THE BLACK ZONE.

sink to a quiet death in the cold northern waters, than to bask here for a time in fancied security in this pseudo-tropic warmth. With every patch of weed—less in extent than an opened hand—a tiny cosmos of creatures kept faith, the faith of unconscious heritage. It was tragic to see a tiny fish or a crab clinging to a thin strand, with no hope beyond another week, the sargassum even now beginning to blacken and water-log. We caught seahorses with astonishing powers of color change, turning quite black at night and pale yellow-orange in the daytime.

The small people of the surface were seldom by themselves; if they were not in schools, then they haunted the bits of weed, or chummed with jellyfish. Great pulsating *Cyanea* jellies throbbed slowly along, umbrellaing with graceful heaves of their massive amber bodies. Behind them trailed for yards the medusa tangle of poisonous, stinging tentacles, and in and out of this living maze of nettles, small fish swam. They were young and inexperienced and they gave me the same sensation as I once had when I saw combat patrols crawl through a snarl of barbed-wire into No Man's Land, where at any moment a Very light might shed its death ray upon them. I watched many of these small butterfish swimming carelessly along, protected from all outside dangers, while every now and then a small entangled corpse showed where the great jelly had taken toll of its pensioners.

Although the weed was so shredded and patchy, yet almost all its usual habitués were to be found, pipefish, sea-horses, filefish and *Pterophryne,* the latter magicked from weed to fish with scarcely any alteration of color, blemishes, berries and fronds.

A host of other surface persons came to our nets, but I will mention only two more. A few Portuguese men-of-war had drifted hither from far-off tropical waters, still iridescent as opals, buoyant as balloons, and among their terrible, fire-searing tentacles, there also swam small fish—fairy *Nomeus* to whom color was as balls to a juggler—one moment banded with black, the next monochrome silver.

Almost the only being who was independent of weed or jelly or the society of its fellows was a little triggerfish, who outcolored even *Nomeus.* Isolated amid this vast waste of waters, this midget would be seen progressing sturdily and unafraid. He was the despair of the artist. Swimming quietly in mid-ocean or in an aquarium, he showed the usual oceanic coloring—ultramarine above, silvery white beneath. As the *Arcturus* bore down upon one of these diminutive triggers, or the face of the artist approached the glass behind which he hung poised, he became purpley suspicious. Another emotion induced a pale green cast, while darkness impelled him to lower the black drop, until he reflected the colors of this printed page. At times (but I am certain never through fear) he turned a strong saffron yellow, while at the approach of death, as weakness seized upon fins and gills, the little spectrum palette of his body was slowly dimmed, and a veil of silvery grey drawn

over all his scales. Through every pigmental vis-
cissitude, every colorful emotion, only his golden
eye and scarlet tail remained unchanged. This
little Joseph of the sea was one of my greatest delights—and in his scant two inches I saw and re-
spected what to me typified fearlessness, dignity,
poise, adaptation, besides incredibly kaleidoscopic
beauty.

I have said that the sea stretched unbroken to
the horizon, but after we had floated quietly
throughout the first day, this was not strictly true.
After dinner I went up on the flying bridge as
usual to watch the sunset, which, however, was
wholly drowned in horizon mist. We had no wake,
of course, as our engines were still, but broadside
on, to windward, which was south-east, was a long
and irregular trail, marking our slow, wind-pushed,
crabwise movement. Slick after slick marked the
places where the galley had poured out gravy or
the engine room oil, and here were gathered a host
of stormy petrels. At sunset there were two hundred and eighty-six and more were coming every
minute. I watched carefully and saw eight
Mother Carey's chickens arrive singly upwind, appearing far away on the leeward side of the *Arcturus* where they could not possibly have seen the
oily slicks. Later three flew into vision at right
angles to the wind, turning only when they were
close. It seemed to me that these little birds, with
their sharp eyes and long, tubular nostrils, probably make use of both senses under different conditions, in discovering, and directing their course

toward, a source of food such as this—doubtless
getting a faint aroma of the floating débris from a
long way down wind, or, on the other hand, perceiving and instantly interpreting any focussed activity
or unusually directed flight on the part of a distant
fellow bird, when upwind or far off to one side.

The mist on the horizon rose gradually after
sunset and smudged out first one star after another until there was only a handful overhead in
the neck of the mist.   This cloudiness presaged a
good night for plankton—for all the floating organisms which love the darkness are kept down far
below the surface by the rays of light from both
sun and moon, apparently as unable to face the
light waves as if they were a rain of venomous fiery
arrows.

I had the gangway put down after eight in the
evening, and with a cluster of electric lights focussed on the water sought to learn something of
the surface night life haunting the darkness here
thirty leagues from Broadway.   It is a curious
thing that while the creatures which swim on the
surface at this time hate the light, yet when they
come within the influence of a focussed search-
light, or any beam of great concentration and
strength, they are unable to resist it, there is
aroused a reaction of fascination, and instead of
fleeing they are compelled to enter its circle and
swim back and forth in the glare of its influence.
The first to come were the squids, but any hypnotic
force which may have drawn them hither became
subordinated to their ravenous hunger when prey

came within sight. On this night all were of a size, about a foot long with a single individual twice that length. They shot back and forth across the circle of light, now scarlet, now pale rose, now white, and when we scooped them up in nets and transferred them to our big tanks neither their activity nor their shift of kaleidoscopic colors ever ceased. Once, and once only there came to the light a great silver-armored, fang-jawed snake mackerel, headed straight for the squids. Instantly, the keen eyes of these mollusks perceived him, their bodies became colorless and they melted into the blackness of the nocturnal sea.

After lunch we made ready to raise our nets, which for hours had been drawn slowly through the black, frigid depths of the Hudson gorge. This lunch, by the way, was an unusually delicious one of fried shark. No officer or seaman would share it with us, giving us thought concerning the human logic of refusing this, and yet with corresponding readiness consuming raw oysters and fried pork!

Up came the nets, sagging heavily, loaded to the very limits of their breaking point. At first glance they seemed filled with a bushel of glass or solid water. A wild thought of submarine ice came to mind and instantly resolved into absurdity, and the moment the first net reached the rail the truth was evident. Our nets had passed through a zone of almost solid jelly composed of untold myriads of salpæ of three species. The tubsful of salpæ on deck increased until our containers were all overflowing. These curious beings consisted of small, angular, double-pointed bits of glassy jelly, each with a pink nucleus, many connected so tenaciously in chains that they could be lifted up like a string of living pearls.

One of the officers with the memory of the shark steak still vivid, said, "Well, I suppose you people would even eat that stuff!" whereat we all solemnly proceeded to eat a salpa. We got no enjoyment from this bit of bravado—just a sensation of very salty hard jelly. And then I aroused all the conventional, anti-Darwinian beliefs of our good skipper by informing him that in eating salpa I had, rather indirectly, been guilty of cannibalism, in that, far from being related to jellyfish, these oblong, glassy blobs of life claimed cousinship with ourselves and other backboned animals. But they have fallen to the lowest point in the scale—even the sea-squirts clinging to our wharf piles parading more highly developed offspring.

Salpæ have an intricate succession of alternate generations, so complex that no genealogist could ever straighten it out. The young larva develops attached to the blood system of the parent and after a while swims off by itself, wholly unlike its parent in appearance, structure and habits, and even quite sexless. After swimming for a time it develops a stolon on which buds form which in time become adult sexual salpæ. These are liberated in sets of long chains, which in turn swim off chummily together, ultimately separating into individuals, who become the parents of the larvæ which complete the cycle.

It looked as first as though we should have to imagine the old Hudson canyon filled with dilute jelly, but on sorting over the hosts of salpæ the more interesting creatures of the deep began to appear. Although in the short time at my disposal I was able to make only a few hauls, yet in this Hudson River gorge I took thirty-two kinds of deep sea fish, some of which are new to science. These were represented by seven hundred and sixty-eight individuals. The most abundant were the delicate little *Cyclothone*—pale ones living in abundance at three to four hundred fathoms, while larger black species were more abundant from five to nine hundred fathoms. They were as delicate as tissue paper, with series of lights along the body and relatively enormous mouths with which they engulfed the tiniest of swimming creatures. When they came up they looked like minute bits of string stuck to the nets, but floated gently out in water all their exquisite structure and illuminating apparatus became visible.

From four hundred fathoms down we secured deep water forms of the Myctophid fishlets which we took at the surface after dark. Some had gloriously brilliant gill-covers, with the eyes scarlet or green. In the lower mid-depths appeared the curious, elongate *Chauliodus* and *Stomias,* with glistening scales, huge mouths and enormously long teeth. Blue-eyed flounders came up, packed safely among the salpæ, and eels never seen at the surface or in any light of day. Some of these were sturdily built, with smooth skin of glistening

bronze, and long, straight jaws which boded ill for
lesser fish which swam within striking distance.
Then there were spectral eels which seemed more
suitable adornments of a fairy tale—inmates perhaps of deep pools beyond Mluna,—pale, slender
eel wraiths, with inconceivably evanescent fins,
large staring eyes, and the most absurd and useless jaws imaginable. With lamentable belittling,
some ichthyologist has named them *Nemichthys*—
snipe eels—the value of this simile being exactly
one-half of one per cent. These remarkable jaws
are thread-like, and just in front of the head they
begin to diverge, each curving away from the other
and ending in a conspicuous round ball. If tentacles were needed by this eel why in the name of
holy natural selection must the jaws be thus sacrificed! These eels were always quite dead when
I found them in the heart of the salpa mass, and
how they live and move and satisfy their appetites
in the icy blackness half a mile beneath our keel
I shall perhaps never know.

Close together in one net were a scarlet and
wine-colored scorpion fish, all abristle with needle
spines on fins and head and gill-covers, together
with a lantern fish with glowing green eyes. Three
other fish which I found living here within thirty
leagues of New York City are typical of the depths
of all the seas in the world. One has been appropriately named *Argyropelecus*—the silvery hatchet—and when young these fish look like nothing
else. They are deep and narrow, with eyes that
stare forever upward, the scales shining silver and

interspersed with groups of luminous lamps. Another related form has the tail end of the body raised high while the skeleton remains where it would in a more normally shaped fish—the fin supports being thus clearly visible and actually outside the opaque part of the body.

Small and jet black spots were occasionally seen embedded in the glassy piles, and in a dish of water each resolved into a diminutive sea-devil, usually a huge mouth with merely enough tail to propel it through the water, or another with a long-armed luminous candle waving about as living bait over the great maw, or again an inch of fish with such elongated fins that it could never have touched the bottom without injury, or in fact have come near anything more solid than the icy water in which it was born, lived and died.

CHAPTER XVI

LOG OF THE ARCTURUS

BY WILLIAM BEEBE AND RUTH ROSE

WE have thought it worth while to present a very brief résumé of each day on the trip, together with the noon position. No attempt has been made to list the hauls or to tabulate any data which belongs more appropriately to future technical papers which will appear in the Zoological Society's ZOOLOGICA.

Up to the present time about twenty of the drift bottles thrown overboard have been recovered, and their distances and time of drift recorded. A typical example is bottle Number 885 which was thrown overboard from the *Arcturus* on June 29th, in N. Lat. 14° 10′ and W. Long. 76° 43′. Fifty-nine days later it was picked up on the shore of St. George Bay, British Honduras, in N. Lat. 17° 33′ and W. Long. 88° 05′, having thus floated and drifted a distance of 720 miles, or over 12 miles a day.

Feb. 10th, 1925. Brooklyn, New York. Sailing day. Reception on board for all our friends to come and look their last. Large crowd. Reporters much perplexed trying to distinguish visitors from expeditionists. Sailed about 2:30; beautiful day till we reached lower bay, then heavy fog shut down and we had to anchor there for the night, waiting for clear weather to swing our compasses.

Feb. 11th and 12th. Progressing slowly toward Newport News to coal for first leg of voyage. Gale and heavy seas.

Feb. 13th. Arrived Newport News at daylight. Coaling begun at once and those of the party who came by train were found to have been waiting for us since the day before. Last day ashore spent in feverishly buying everything in sight, for fear we might have overlooked some necessity. Ten-cent stores practically cleaned out. Everyone obsessed by feeling that this was the last chance we should ever have to purchase anything.

Feb. 14th. Sailed at 4:30 p. m., headed for Bermuda.

Feb. 15th and 16th. Heavy seas. Ship rolling deeply, and several wan faces show the effect. Attendance at meals spasmodic. Everyone busy unpacking and dragging quantities of things out of the hold, to distribute in their proper places. Not possible to arrange things very neatly, as ship is far too active and every object falls down or slides around.

Herring gulls followed us up to the evening of the 16th, and then left in a body. They alighted on the water alongside again and again, occasionally feeding, but more often only sitting quietly, very evidently resting. Even when the waves were highest and the wind strongest, they rested thus for five minutes at a time, if not much longer.

Two parasitic jaegers were about most of the day, flying somewhat more easily than the gulls. The projecting central rectrices were plainly visible. They went through a regular routine of flying well ahead of the vessel, alighting and resting until the ship just passed them, when they rose and again flew ahead.

The gulls (twenty-two in all) spent most of the time, when not resting on the water, balancing about twenty to forty feet above our heads, headed up-wind just to windward and almost over the ship.

Feb. 17th. Fairly quiet sea and beautiful day. Swung the pulpit over the bow and had our first trial at catching weed from it. Caught a tubful of weed but not a single fish in it, and only moderate numbers of shrimp and crabs. The weed all in small patches or smaller bits and in long lines running with the wind, at right angles to our course. Much of it rather old, with distinct new growths at the tips. The government's map of the Sargasso Sea for February shows Bermuda well clear of the Sargasso area, but our experience of to-day shows this is quite wrong, or that this is an exceptional year.

Not a bird or cetacean all day. A bumble bee came aboard, having flown four hundred miles from the land.

Noon position: Lat. 33° 27′ N; Long. 68° 31′ W.

Feb. 18th. Weed less abundant but in larger masses, and much younger (lighter yellow) in appearance. No fish seen beneath it, and that which we scooped up yielded only shrimp and crabs. Saw one flyingfish. No birds, whales or other life. Bermuda sighted about noon. Picked up black pilot about 4:30, and he took us into St. George's, through the extraordinary channel, where the *Arcturus* seemed to scrape the coral cliffs on either side. The water is so clear that the reefs show up alarmingly. Scrubby cedar trees were almost within plucking distance, growing among rocks and sand, and sheltering little negro cabins half the size of our deck-houses. Tiny islets dotted the quiet harbor, and as the brilliant sunset faded, small sounds from the town came out to us with fine-drawn clearness. Scudding showers during the evening settled into an all-night deluge, but we caught a number of small fish, a crab (*Callinectes or-*

natus), and a big swimming shrimp (*Penaeus braziliensis*) from the gangway.

Feb. 19. Up and ashore in torrents of rain. A lovely town, with winding walled streets, and gates standing ajar on glimpses of gardens crowded with flowers and the sea for background, or casual moments in the family life of the black Bermudian. We all succumbed to another spasm of ultimate shopping, though there was little enough to buy, and at 10:30 returned to ship with such loot as could be had, one item being a ship's cat, carelessly overlooked in New York. At 11 we steamed out through the hole in the wall, and rolled all day through a grey sea, aiming for 30° North and 60° West, as that is well within the area of the Sargasso as mapped.

The rain stopped as soon as we got away from the island. Pulpit lowered and two tubfuls of weed caught, but not a single fish. Quantities of crabs, mostly *Planes minutus*, and a single *Portunus sayi*. Many shrimp, chiefly *Leander tenuicornis*, and a few *Latractes fucorum*. One of the latter had a full-grown parasitic *Bopyrus* in its left gill. If gathered together, the weed we have passed in the last three or four days would make vast plains, and yet it is all in small heads or strands, and with little life on it.

Feb. 20th. At dawn Captain blew foghorn to call attention to tropicbird or, as he called it, "marlinspike," flying round ship. Large empty ocean, ship rolling too much for microscopic work, though weed was scooped up, despite the seas. Sounding machine and trawls still to be put in working shape. Small patches of nearly lifeless weed all day.

Noon position: Lat. 32° 22′ N; Long. 64° 39′ W.

Feb. 21st. The foghorn at dawn announced the tropicbird again. About noon it looked as though we had reached the Sargasso frontier at last. Comparatively large patches of weed, two or three yards across, strewed the sea at intervals as far as we could see in all directions. However, we soon ran out of this area. A school of dolphins stayed with us for a long time, and we began to see flyingfish. Caught a little weed but found it as barren as ever,—no fish or nudibranchs, and only small crabs. Grey weather continues, and a fair amount of roling.

Noon position; Lat. 29° 43′ N; Long. 59° 38′ W.

Feb. 22nd. Tropicbird at dawn again. No Sargasso Sea. Too rough for trawling. Divided day into half-hour watches in bow, but only flyingfish seen. Some of the staff still suffering from constant rolling. Everyone getting nervous for lack of weed and work.

Noon position: Lat. 27° 49′ N; Long. 57° 45′ W.

February 23rd. Two tropicbirds at dawn. Two blasts from foghorn. Sea quieter but heavy ground swell. Made our first Station.

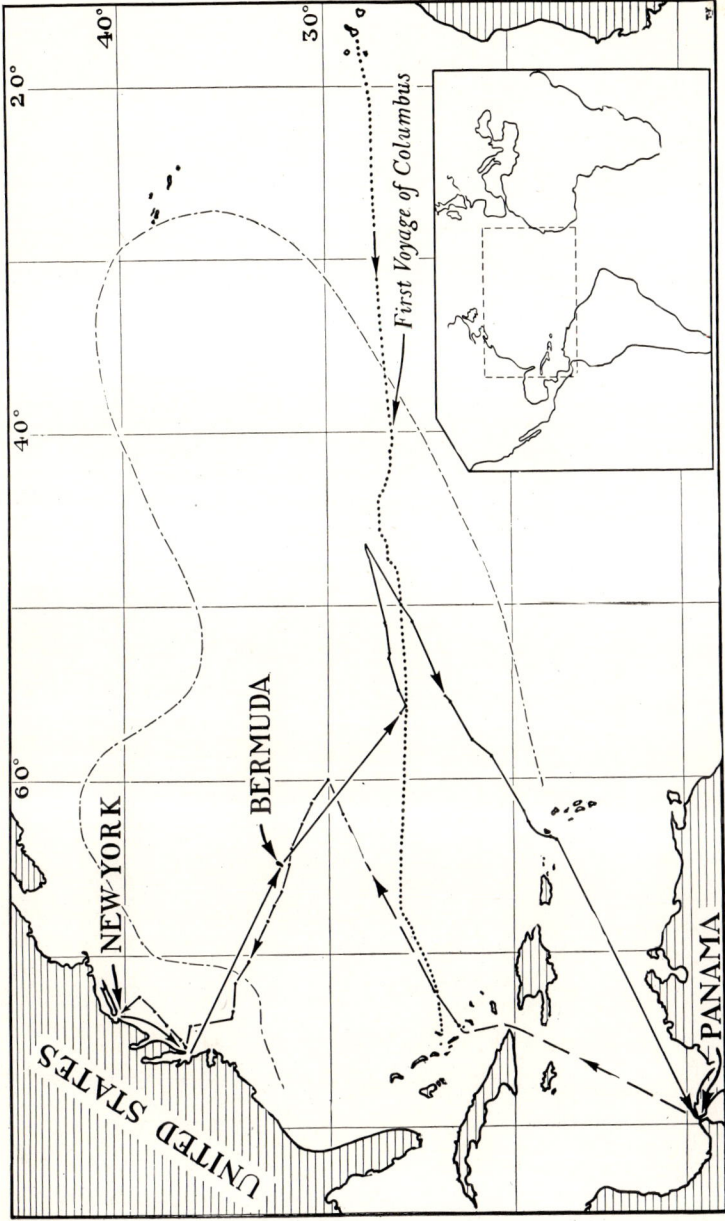

First Voyage of Columbus

UNITED STATES

NEW YORK

BERMUDA

PANAMA

Fig. 61.—Atlantic Routes of the First Voyage of Christopher Columbus and of the *Arcturus*.
The Limits of the Sargasso Sea are indicated by the fine, broken line.

Fig. 63.—First Officer McLaughlin.

Fig. 62.—Captain James Howes.

Put over two Petersen trawls from stern and lost both. Third net towed at two knots, and brought in one half-inch squid. Put the boom-walk into commission and towed silk surface nets from there, catching quantities of weed from the small pieces that floated past. The usual small amount of life,—crabs, shrimps and nudibranchs, and some interesting egg clusters, probably molluscan. No fish at all. From the deck we saw occasional large swimming crabs, and wind-rows of loose berries, all indicating a complete destruction of any masses of sargassum which may have been consolidated at one time. Weed fairly abundant, though so much scattered, and young.

Chiriqui broke out of his cage and wrecked one of the rooms before he was detected and captured.

At 8 p.m. made surface hauls with half-metre and one-foot silk nets and got exciting results. Thousands of organisms of all kinds, tunicates, larval fish, including twenty Leptocephalus eel larvae, feather-tailed Copepods, Pteropods and Heteropods, sea-worms,—in fact a solid month of investigation would not exhaust this one haul. The Leptocephalus were ghostly transparent, except for the solid circle of mother-of-pearl of the eyes, which glowed like fire. The nearest of any example in books is pictured in *Depths of the Ocean,* page 92.

Noon position: Lat. 26° 06′ N; Long. 55° 56′ W.

Feb. 24th. The tropicbirds were prompt and so was the foghorn. We all think Nature is wonderful, but some of us wish birds would sleep late. They probably follow the ship at night and go off in the daytime, but these are assuredly the same pair, as one has lost a long rectrice and is easily recognizable. Two nets put over at 4 a.m., and more Leptocephalus caught. At 6 a.m. another net brought in a good haul, though it was daylight. A kittiwake came aboard later, black with oil, so that we thought it was a new species until it came close and we saw that the black breast was the result of following a tanker. On wireless advice from other ships in vicinity, we changed course slightly, as they report more weed further east. Sounding machine in difficulty, so we wallowed in the trough of the waves for hours while it was repaired.

Noon position; Lat. 26° 17′ N; Long. 55° 09′ W.

Feb. 25th. Tropicbirds showed great discretion, not appearing until seven o'clock. Made one haul with Petersen trawl, and two with vertical nets. Remarkable colonial Siphonophore and an unusual squid were the most interesting specimens. Much discussion of Museum groups, and a rough design set up in library to work on. Sewing machine busy making nets against the time when we shall be really busy.

Noon position; Lat. 26° 42′ N; Long. 53° 11′ W.

Feb. 26th. Day dawned with a beautiful tropic sky, the bluest water in the world and a heaving swell with mighty waves, thirty feet from crest to crest. The big drum was put in commission and a Petersen trawl with mouth 3 x 7 feet put over to windward over the starboard rail. The wire ran out evenly and so smoothly that we were tempted to increase the speed, and ran out 4000 metres in sixty-six minutes. From 11 A.M. to 2:30 P.M. we barely maintained steerage way, running the port engine at slow speed.

The intake of wire was at half the speed of the paying out. 3000 metres had come in and all seemed to be going well, when a great cry of alarm broke from everyone, as from the blue depths a great mass of tangled wire rose steadily toward the cleeve at the tip of the outswung boom. The drum was stopped just in time to prevent the wreck that would have resulted if that snarl of cable had gone ten feet further. The first officer and four deck hands performed miracles of cutting and splicing, so that the trawl was saved. That gigantic cluster of festooned cable coming over the rail was the nearest thing we've seen to a giant octopus. After all, however, only a hundred metres of cable were damaged.

The end of the trawl was placed carefully in a tub of water, the bag untied and the net opened. To a Gloucester fisherman accustomed to seeing his net bulge with hundreds of food-size fish, this haul would have been a complete and perfect failure, and at first we thought so too. Then in the brown meshes, we began to see glistening silver patches, a streak of black floated off the net and became a strange fish, a scarlet blot was unfolded from a hidden corner of the trawl and almost invisible transparent creatures were betrayed by a glint of light along their paper-thin bodies. The first live fish of the expedition was in this haul, a tiny globular chap, energetic but short-lived. He was christened "Zoop," after our German mess-boy's pronunciation of the first course at dinner. The complete list of the haul follows:

STATION 7, PT1

    1 Wing-finned Globe-fish
    3 Sternoptyx diaphana
    7 Cyclothone signata
    1 pink-banded fish
    1 small black fish
    I squid, 15 inches long
    2 Siphonophores
    2 scarlet shrimps
    1 Medusa—Periphylla
    5 Coral pink Sagittæ
    1 transparent Sagitta
    Many Salpæ

Noon position; Lat. 26° 57′ N; Long. 51° 14′ W.

Feb. 27th. Tropicbirds are distinctly a drug on the market, there being a supply of two, and absolutely no demand. Fair and warmer, real blue sea, and the smoothest we have had so far. No stops to-day, as we are headed for the Atlantic Ridge, where we shall loiter for some time, weather permitting. Charlie Fish gave us a résumé of plankton work in the evening.

Noon position; Lat. 26° 55′ N; Long. 49° 13′ W.

February 28th. Station 9 on the Atlantic Ridge. A sounding in 2400 fathoms. The bottom sample was Globigerina Ooze, which showed under the microscope as tiny round, reticulated balls scattered widely in the amorphous particles of mud. Then a series of temperatures and water samples for salinity. Next a vertical haul, which did not amount to much so far as quantity of specimens was concerned. Then five metre nets were put down at depths from 500 to 3000 metres. By that time dinner was ready, so they were left to tow until 6:30. The upper one had twisted on the way down, so the aperture was shut fast, the next two had vanished from the cable entirely, the fourth came right to the surface, almost within reach, then pulled off and sank. The swivels of the lost three had broken or worn through, being too light for this work. The fifth contained a rather good assortment of creatures.

We are half-way between Africa and America.

Noon position; Lat. 27° 50′ N; Long. 46° 58′ W.

March 1st. Our first really perfect day. Everyone went swimming in mid-Atlantic. Much other active outboard work done. Identification and study of specimens already caught occupied the rest of the day.

Noon position; Lat. 27° 58′ N; Long. 46° 54′ W.

March 2nd. All day spent in putting down bottom dredge, and bringing it in again. More than three miles of cable let out. Towed for about an hour, and at 5 P. M. dredge reached the deck once more. During the towing there was tremendous jerk that actually stopped the ship and made us thankful for the automatic towing device that undoubtedly saved most of the gear from being torn away. Front bar of dredge bent almost to a semicircle on the obstruction, whatever it was, and among other contents a beautiful glass sponge and large pieces of black lava.

Noon position; Lat. 27° 53′ N; Long. 46° 24′ W.

March 3rd. Lovely day; able to put over small boats for first time. Did surface collecting from them. A very rusty Welsh tramp steamer was sighted about five P.M. She was west-bound but altered her course slightly to cross our bows within a hundred yards. Presume she was curious to see why we were drifting aimlessly about in mid-ocean. She looked a wreck, so we feel we have seen one Sargasso derelict after all.

Took soundings and when the wire came up we detached several feet of curious red tentacle or tissue, with strands of colorless, sticky stuff which might have been bits of some large Siphonophore.

From 7 to 9:30 P.M. towed three surface nets astern and took;

142 Leptocephalus
Large purple many-spined Zoea of an Astacus
3 small squid, all very different
3 Pipefish, one with covered ova in its pouch
More than 40 large and small Sagittæ
Several small fish
1 red-eyed Annelid
Many Pteropods of several species
Untold thousands of black Candace copepods and the bright steel-
blue Pontella

Noon position; Lat. 27° 57′ N; Long. 46° 42′ W.

March 4th. Decided that the Sargasso Sea was too disturbed for useful study and turned toward Panama and the Pacific, planning to come
back here in the summer, when we hope the weather will be more propitious. And, having given up the Sargasso Sea, we are now passing
more weed than at any other time! Tubful after tubful scooped up
today, and in it found our first *Pterophryne* of the trip. Nudibranchs
particularly plentiful and many pipefish. Half a dozen boobies seen in
distance, and flyingfish numerous. Sighted northbound steamer on
horizon.

Noon position; Lat. 26° 43′ N; Long. 48° 52′ W.

March 5th. Fairly rough, partly cloudy. Sounded at 7:30 in 1983
fathoms, bringing up red clay and Globigerina. At 9 A.M. took
temperatures, then sent down two Petersen trawls to 500 and 250
fathoms. Most successful hauls so far, bringing in hosts of giant-
mouthed Cyclothones, many silver *Argyropelecus,* and best of all, a
tiny stalk-eyed fish, *Stylophthalmus,* so delicate that we were afraid to
touch it until we found that it was not quite so fragile as it looked.
A third trawl was put over, and then as the crew and all the men had
worked all day, the women of the party volunteered to take over the
tedious work of oiling, beating and wiping the cable as it came in.
Another good haul. No one left the laboratory and main deck before
midnight.

March 6th. Ship rolled wildly all night, to an accompaniment of
dismal crashes. Nervous scientists staggering about wet decks to
see whether said crashes were irreplaceable laboratory equipment or
merely kitchen supplies. No great damage done. Wallowing along in
the trades now, swell so heavy that course has been changed to take
it head on. No one has the heart to suggest trawling. Working
on captured fish so far as the motion of the ship will permit.

Noon position; Lat. 25° 14′ N; Long. 52° 54′ W.

March 7th. Squally, rainy and rough. Put over Petersen trawl
at 9 a. m. in 250 fathoms. Towed four hours. Result:

1 pygmy sailfish
13 Cyclothone

4 Valenciennelus
3 Argyropelecus
1 large red shrimp with antennæ 145 mm. long, tip luminous
2 small red shrimps
6 pale grey shrimps, with slight pinkish tinge
Many Sagittæ

These fish were less distorted than any we have caught before. Noon position; Lat. 23° 42′ N; Long. 55° 09′ W.

March 8th. Sunny, warm day till about two p. m., then cloudy, wind and rain. No trawling or dredging. Worked on yesterday's specimens. A surface net in the evening got two specimens of the blackbeard fish, one of them still alive, and a transparent flounder (*Bothus atlanticus*) with the eyes symmetrical.
Noon position; Lat. 22° 23′ N; Long. 57° 16′ W.

March 9th. On Echo Bank this morning, sounded in 2000 fathoms, which is much more shallow than the general depths given hereabouts. No hint of the 34-fathom depth marked P.D. on charts, though first officer took many soundings with the 100-fathom ship's line. Put over fifty-foot otter trawl to 250 fathoms and towed it for three hours. Brought it in last few yards with donkey-engine and the whole net ripped off just as it came alongside. Put another one over at once and towed two hours. Got little new and not many specimens we desire. Sea rose and tremendous swells came down from the north. At 11 p. m. shaft of circulation pump broke, so engines stopped and wallowing began.

March 10th. No one slept last night. We drifted, rolling to the bulwarks until 5 a. m., to the music of breaking dishes and hammering from the engine-room. Finally by attaching fire-pump we got under way and made slow progress. Pale wan faces at breakfast table, after a night of defying gravity to stay in bunks. No work possible all day. Only place to sit is flat on the deck and even then we skate to and fro. At luncheon and dinner the *Arcturus* had to be headed north to enable us to get food into our mouths.
Noon position; Lat. 21° 12′ N; Long. 58° 53′ W.

March 11th. Like yesterday. Got five knots out of the engines and rolled unceasingly. Brilliant day. Flyingfish seen all day and considerable weed, some of a new species, with very numerous berries packed with the leaves on very long straight stems. Some work done, by means of sitting on the deck, with feet and back braced, and balancing typewriter, books or canvas on the knees. When over or near Echo Bank, put tow-nets astern and caught two young vertical flounders, a puffer and a young flyingfish, also several megalops of non-swimming crabs. This hints at proximity of land, nearer than two or three miles down. Everyone tired out with the strain of constantly holding on.
Noon position; Lat. 20° 10′ N; Long. 60° 26′ W.

March 12th. Ocean behaving somewhat better to-day, so in after-noon took series of temperatures, and made a vertical haul, which yielded five specimens of Amphioxus, an event rare enough to cause excitement. At night surface net brought in young flounders, flying fish and **Leptocephalus**, also a fish like a baby sea-serpent with a long appendage on his back, and what may be larval forms of the stalk-eye, *Stylophthalmus*. A marvellous moon-bow to-night, a pallid coppery arch like the rainbow's wraith.

Noon position; Lat. 19° 21′ N; Long. 61° 57′ W.

March 13th. Sighted meagre little **Sombrero Island** early this morning, and passing it, spent the afternoon and night in the lee of St. Martin's; the lovely peak of Saba is dim and dream-like in the distance. After dark lowered the gangway and hung powerful lights close to the water, and with hand-nets captured mullets, half-beaks, needle-fish and flyingfish, and a squid, all attracted by the blaze of electricity. All the fish had the typical coloring of the pelagic surface forms,—dark blue above, silvery beneath. The squid was brilliant green and yellow, and vigorously bit **Serge**, his captor. Seven isopods were taken and no sooner did a captured fish turn on its back in the aquarium and show the first signs of distress, than these voracious crustaceans attacked it and literally tore it apart, an interesting example of swimming scavengers so far from land. When we put a light ten feet under water flyingfish flitted past it like moths around a candle.

Noon position; Lat. 28° 17′ N; Long. 62° 28′ W.

March 14th. At dawn we were near Saba, and made for Saba Bank to put over the 40-foot trawl. Depth supposed to be 300 fathoms, but we shall not trust to charts again, for found it was only 45 fathoms. Pulled in trawl immediately and found it un-harmed and filled with sponges, corals, and all sorts of creatures, vertebrate and invertebrate, burr-, porcupine-, and triggerfish, star-fish, anemones, crustaceans, sea-cucumbers, and dozens of smaller animals. Repeated the haul with the coarse rope dredge a number of times and covered the deck with enough coral to build a house. We are supposed to be in the lee of Saba, but the island doesn't seem to possess such a thing, as we rolled too much for comfort. Clouds always cling to the summit of Saba, and there are showers and rainbows coming and going all about. Steaming back to St. Martin's to-night and will then drift down to Saba again.

Noon position; Lat. 17° 40′ N; Long. 63° 20′ W.

March 15th. Fair, strong wind. Close to Saba at daylight. Put over rope dredge in 250 fathoms and got only a few starfish. Sounded, and bottom sample showed many globigerina and a few pteropods. Then moved over to the Bank, and in only eight or ten fathoms made five or six hauls as rapidly as the dredge could go down and come up. Wonderful lot of marine organisms. It would pay to come here and stay a year. Sponges of every form,—carrots with stubby tops and roots, vases, footballs, cups, fans, platters and clubs, with every color in the spectrum represented, in glaring tones and startling contrasts.   There were scores of marvellous serpent starfish, no two alike, and worms, mollusks, shrimps, and urchins. Now and then we came across beautiful little fish, some defying attempts at identification,—scarlet, yellow, striped, spotted, many with five finger-like pectoral rays, specialized for clinging to the coral as Pterophryne's fin-rays are for holding to the weed.

Someone saw a fish's head protrude for an instant from the centre of a giant scarlet-tentacled holothurian, and presently out crept a wonderful *Fierasfer,* bright iridescent copper and green and gold and red, with a long eel-like body and extraordinary eyes, a fish that spends its life in the strange sanctuary where we first saw it, not parasitizing the holothurian, except in so far as shelter is concerned.

Noon position; Lat. 17° 39′ N; Long. 63° 16′ W.

March 16th.   Running at half-speed with wind, current and sea behind us, we are making seven and a half to eight knots across the Caribbean.   No stop.   Passed a school. of blackfish or grampuses, playing or courting.   Big flippers spanking the water resoundingly. Two lying quietly at the surface side by side.

Noon position; Lat. 16° 35′ N; Long. 65° 32′ W.

March 17th.   For most of the day a school of tunnies darted to and fro just under and ahead of the lowered pulpit.   Underwater they look brilliant violet, with all fins golden yellow, except the caudal which is black.

March 18th.   Fair; strong following wind.   Two Coryphæna or dolphin-fish caught from stern by crew, on a No. 3 cod-hook and piece of white canvas, while the ship was making eight knots.   One weighed nineteen, the other twenty-three pounds, both females in full breeding condition.   Got many parasites from mouths and under skin.   Seems to be no truth in poetic rumor that dying dolphin flushes with rainbow changes of color, as the only thing that happened to these was that one developed a line of dark vertical bars which the other lacked altogether.   Flyingfish very abundant, and weed passing in small patches. Several Physalia seen.

Noon position: Lat. 14° 03′ N: Long. 71° 21′ W.

March 19th and 20th.   Higher sea, but we are steady since wind, current and swell are following.   Thousands upon thousands of flyingfish around us all day, from the size of a bee to those four or five inches long.

March 21st.   Entered Limon Bay this afternoon and anchored as near Fort Sherman as we could, planning another collecting trip such as we had at Devil's Hole three years ago.   Everyone ashore in evening.

March 24th. Spent day at Devils' Hole, acquiring many specimens,—snakes and birds as well as marine creatures.

March 28th. All repairs on ice-machine, and pump finished, more
coal taken on, and four more expeditionists acquired, we went through
the Canal to-day, and entered the Pacific at 11 P.M.

March 29th. In the Pacific, an ocean that can apparently be depended on,—at least this part of it. There is hardly a ripple on the
surface, and the sun shines and the temperature is perfection. Everyone busy cleaning laboratory, settling new quarters and generally
clearing the decks for action. Put out surface nets in evening, and
in an hour brought in solid quarts of plankton.
Noon position; Lat. 7° 15′ N; Long. 79° 56′ W.

March 30th. Up early, taking sounding in 2070 fathoms, followed
by Petersen and otter trawls, and silk nets at varying depths, while
some of us collected with dipnets from boom and pulpit. Every
net comes in bulging with so many organisms that the mere preserving and recording of the creatures requires hours, to say nothing
of studying them. Just before dinner the water became thick with
innumerable jellyfish and similar planktonic animals, so that from the
upper deck the ocean alongside seemed to be clouded and milky.
We tried raw plankton as a food, and found that so far as flavor is
concerned, one might as well take salt water.
Noon position; Lat. 5° 03′ N; Long. 81° 08′ W.

March 31st. Station 28, 5:15 to 10:30 a. m. Sounded in 1805 fathoms.
Put down otter trawl 500 metres and three surface metre nets.
Captured *Argyropelecus* and *Cyclothone* in the trawl, and *Polynemis*
and young flyingfish at the surface.
Station 29, 3:50 to 4:20 p. m. Put down a Petersen to 500 metres
and the silk net to 250 metres, obtaining *Cyclothone, Argyropelecus,
Myctophum,* and several species of flounder.
Station 30, 8:00 to 8:20 P.M. Two surface nets. Results, young
Coryphæna and flyingfish, Myctophids, and quantities of inverte-
brates.
Noon position; Lat. 3° 35′ N; Long. 83° 01′ W.

April 1st. Sounded at 5:30 in 1826 fathoms. Hard bottom, nothing
in bottom samples. An extraordinary sight greeted us at dawn. As
far as the eye could see stretched a clearly marked line of foam,
zigzagging to the horizon in a NE and SW direction. On the south
side of the line the water showed dark and rough, while to the north
it was lighter and smoother. We later found that the temperature
of the smoother water was 2° lower than that on the southern side.
This line, that wound across the placid sea like a river meandering
through smooth fields, marked the meeting-point of two great currents,
and within its narrow limits swam or drifted or flew an amazing
quantity of varied life. Boobies, petrels, phalaropes, gulls, tropic-

birds, and frigatebirds dived for the abundant food that was concentrated here, or rested, fullfed and lazy, on the water. Numbers of logs and pieces of timber bore each a row of gorged boobies, and the phalaropes that flushed before the slowly-moving *Arcturus* followed in their flight the curves of the current rip as carefully as though it had been a cleared trail through a forest. A school of five or six hundred dolphins leaped and played to and fro across the line, and the blue and silver of a myriad flyingfish flickered everywhere. Great patches of the sea were colored deep purple by countless millions of tiny Salpæ, and every drop of water held a bright little Copepod. Under the shelter of the floating logs lurked fish, feeding on the worms and crabs that covered the rotting wood, and larger fishes, such as the gleaming Coryphæna, in turn fed on them. Several of these logs were hoisted on deck and from them fish and invertebrates were taken. Two large turtles drifted peacefully past, and a little later, a sea-snake was scooped up from the boom-walk. We put over small boats and rowed about, catching pelagic anemones, Porpita, Glaucus, Halobates, balls of mollusk eggs, Ianthinas, beautiful white-winged flyingfish and hosts of crustaceans. There were always sharks tacking about the ship, and one was harpooned. The *Arcturus* drifting along the rip to-night, seemingly magnetized and held to it as completely as one of the logs.

Noon position; 2° 26′ N; 85° 23′ W.

April 2nd. Still drifting in the rip. In the night we heard breakers, a startling sound till we realized it was noise of the spouting white-caps that mark this zoölogists' paradise. During the night we drifted eleven miles to the westward and turned completely around twice, but never left the rip. At 8 a. m. put over the Petersen trawl but the conflicting currents threw it about until it went under the stern and the bridle caught on a propeller blade. The boatswain and a sailor cut it and after re-rigging the trawl, we towed for an hour but got only a mass of salpæ and pteropods. While it was out, we steamed slowly west along the rip and every moment was exciting. Caught two more sea-snakes and a 32-pound Coryphæna which had been feeding on paper nautilus, shells and all. A twelve-foot hammerhead shark stayed with us for some time but refused to be either harpooned or baited. Several deep-sea hauls proved unprofitable, bringing in only a few specimens of Cyclothone from 500 meters. This is decidedly a surface region. Night collecting from the gangway under bunch-lights is a weird feeling; the black water surrounding the small circle of intense light, the roll of the ship which, slight though it is, seems much greater under these circumstances and which throws the water into a turmoil now and then; the vaguely-seen forms gliding in the depths or suddenly taking shape as they come swiftly toward the surface, and the difficulty of judging distance with the lights altering perspective make the proceeding rather dream-like. Squids are very numerous and of every size from three inches to three or four feet. One huge one shot out of the water twice, with tentacles reaching at the gangway; it must have been at least eight feet long, with enormous staring eyes like great pools in the pale flesh. Paper nautilus floated past, clinging to each others' shells in single file. Creatures that looked like silver dollars came into the circle of light, others which, until lifted in the net, seemed to be nothing but a flat mass of bubbles. Halobates skated on the surface, while crabs large and small paddled from darkness to light and back again. Coryphæna made swift raids on the half-beaks and flyingfish that were attracted by the glare, and once a whole flock of white-winged flyingfish came fluttering into view, with a flight so deceptively like the butterflies they resembled that the swing of the net in mid-air was that of an entomologist. One surprise was a great head of spongy material which proved to be a mass of incalculable millions of Coryphæna eggs. And another was the taking of a large *Sternoptyx*,—larger than any we have taken in the deep nets—floating at the surface and much bitten by crabs.

In the evening we got under way and steamed SW for the Galápagos.

Noon position; Lat. 2° 08′ N; Long. 86° 17′ W.

April 3rd. No need to mention weather in this halcyon region. Passed into the Galápagos zone this morning and set half-hour watches in the bow, each watcher to record every sign of life seen. Coryphænæ, tunny, flyingfish, a shark were the fishes recorded; petrels, boobies, tropicbirds, noddy terns and frigatebirds were seen, the last three seeming to tell of land not too far off, and at 5 p. m. a Galápagos Gull flew over; Halobates were present all day, and the only other insects seen were several moths, but these may have come with us from Panama; the usual surface organisms, such as Porpita, Glaucus and Ianthina were frequent, and one fairly large Pyrosoma drifted by; the only mammals were a school of porpoises in the distance.

In the afternoon a Petersen was put down to 700 fathoms, and the best haul of the trip was secured. Two black sea-devils, with jointed rods springing from the heads, a fish with something that looks like an elephant's proboscis, hundreds of Cyclothones, many *Sternoptyx* in splendid state of preservation, a large prickly deep-sea shrimp, strange-looking red and black jellyfish (*Atolla*) and hosts of small scarlet shrimps and pink Sagittæ.

Late in the evening a half-metre surface net towed for twenty minutes yielded one hundred and forty Myctophidæ, all alive and brightly luminescent. Many of the larger ones are *Myctophum coccoi*.

Noon position: Lat. 0° 43′ N; Long. 88° 34′ W.

April 4th. Sighted Tower Island at dawn to the north. Boobies, frigatebirds and petrels in numbers. Steaming toward Seymour Bay, attended by hundreds of dolphins that converged toward us from all sides. Some of them very large. Sounded at 6 a. m. in 559 fathoms. A small hawkmoth flew past when we were 35 miles off Seymour. Sighted Indefatigable at 8 a. m. The first sea-lion

Fig. 64.—Isabel Cooper, Staff Artist, Painting a Living Fish.

Fig. 65.—Dwight Franklin, Coloring a Plaster Model from a Living Fish.

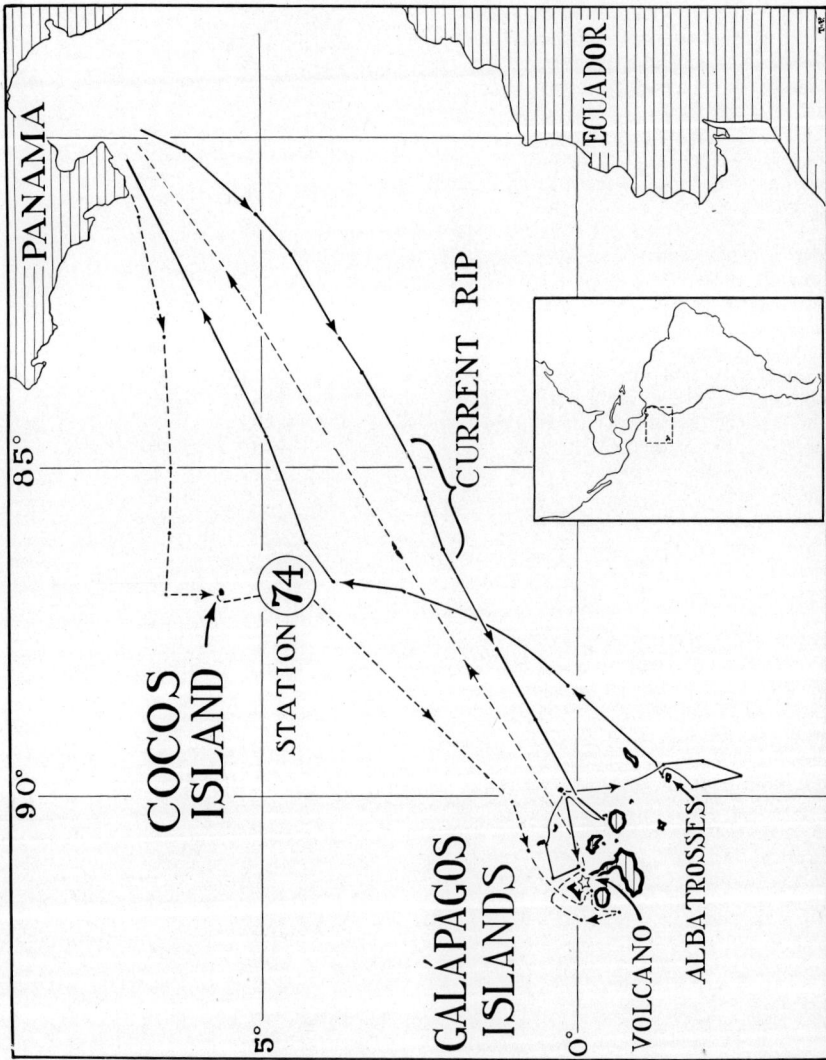

Fig. 66.—Pacific Routes of the *Arcturus*.

The First Trip is indicated by a heavy, solid line.

was seen ten miles from shore. At 10 a. m. sounded in 710 fathoms
and found temperature of water ranging from 82° at surface to 47°
at 500 metres. This latter is six or seven degrees colder than at
the Current Rip, so perhaps we are entering the Humboldt Current.
There were many shearwaters about, one flock of twenty-five fishing
in one spot. Anchored a mile off Conolophus Cove in early afternoon.
Much fishing done, and eighteen or twenty species taken on hooks
from stern and boom-walk during afternoon. In the evening at the
gangway halfbeaks, puffers and so on came to the light, and many
megalops and zoea, as well as stalk-eyed shrimps and copepods were
secured.

April 5th. Everyone ashore and a real home-coming sentiment
strong in those of us who were here three years ago. The place is
as wonderful as ever,—one place in the world that remains unchanged.
Renewing old acquaintance with Conolophus, Amblyrhynchus and
other first families. The pelican colony is small but flourishing, as
one nest contained four eggs. Another nest with two and one with a
single egg. Several unfinished nests. Also found a yellow-crowned
night heron's nest with three eggs, and saw oyster-catchers which
seemed to be breeding. Seining on the beach just as we used to
do, while pelicans, coming home from fishing trips, stooped on the
wing to look at us curiously, not more than arms' length above us
as we splashed and tugged at the nets. Saw a noddy tern alight
on the head of a pelican without eliciting any protest.
    To the astonishment of the old-timers, it rained heartily this morn-
ing from eight to ten.

April 6th. Through the glass-bottomed boats, as we row about the
bay, we can see hosts of fish, blue and yellow, green and red, ming-
ling with less gaudy ones. Rain again this morning. Even in the
Galápagos it is always an unusual season! Most of us swimming for
hours, and seining on beach. An expedition to a beach to the southward brought back four adult skulls of *Pseudorca* and one young
skull, with many bones. Turnstones, black-necked stilts and tattlers
are running along the beaches, while mockingbirds and finches are as
numerous as ever. Goats seen in the distance but they remain the
only wild things on this island.
    Everyone groaning with sunburn to-night.
    Left the Seymour anchorage at midnight under a full moon and
headed for Tower.

April 7th. Dropped anchor in Darwin Bay, Tower, before noon.
Captain much worried about getting in, as he sent out the second
officer in a small boat to take soundings with the hand-line, and one
minute he would shout, "No bottom at fifteen fathoms," and the
next cast would be, "Six feet!" This on the west side of the bay.
At last anchored near our old spot, so close to the east cliffs that it
looks perilous, but isn't. All ashore after lunch, and found the rookeries as populous as ever, with frigatebirds and boobies nesting in

quantities behind the beach, but few gulls and none nesting that we saw. Sea-lions sleeping in the tide-pools, and mockingbirds feeding their young. Walked along shore to the wonderful pools under the cliffs, finding doves' nests with eggs on the way, and collecting the small moths that are abundant in the scrubby growth. Grasshoppers common, and many little Hemiptera on the low green plants. A queer parasitic plant very common here; it is pale straw-color, and looks like the shrivelled remains of some low growth at a little distance; on examination it is a mass of filamentous strands, completely covering some clump of vegetation over which it is heaped like a haystack.

April 8th. Rowing round cliffs to east and south, found hundreds of fork-tailed gulls (*Creagrus furcatus*) and noddy terns in even greater numbers, nesting in cracks and on tiny shelves of rock. There were shearwaters in small flocks of six or eight, and tropicbirds alone or in pairs. A few pelicans were roosting here and there, and many white boobies along the summit of the cliffs. Dwight Franklin explored island to opposite coast, but found little change in the country all the way; coming back he wandered to the northward, and found a crater lake. Did not get back till after midnight. Rain in the afternoon and showers in evening. Nothing like this three years ago. Splendid fishing all round bay and outside, and enough groupers and mackerel being caught to feed all fifty-six of us.

April 9th. Tried diving helmet for first time, and found it most exciting experience. Trite but true to say it opens a new world. Went down several times in about fifteen feet of water, experimenting with pressure, rate of pumping, and so on. A large shark was swimming nearby, but paid no attention to the diver. The strange beauty of the submerged scenery is hard to describe. The range of vision is limited to perhaps forty feet, while everything beyond that is wrapped in a soft, luminous fog, a delicate blue haze like that of a concentrated Indian Summer, in which shadowy forms weave to and fro. There is little coral at this spot; the bottom is covered with pavement-like volcanic rock, like that of which the island is mostly made.

Ice-machine has sprung leaks, so that ammonia fumes make the engine-room almost unlivable. We are madly eating meat and fruit to save it.

April 10th. Staff now divided into parties, with special assignments for each, such as shore collecting, plankton collecting, fishing, identification and dissection of fish caught, painting, photographing, mapping and sounding the bay, and diving in the helmet. Tremendous activity all day. Fishing and diving from small boat under cliffs on southeast side of bay, where pelicans, gulls and terns were nesting. Sea-lions were sleeping in deep crevices, or swimming slowly along shore, and whenever one drifted past, all the large fish vanished instantly. Big hammerhead shark caught in gill-net, and nearly upset the boat when dragged into it. White-striped angelfish floating on their sides at the surface, always just out of reach, and taking no interest in any bait. A bit of orange-peel, tossed into the water, attracted several to nibble tentatively, but as soon as a hook was added, they ceased to bite. Saw them rise to bird excrement as it fell on the surface. *Paranthias furcifer* was the fish most easily taken on the hook, and all round the boat noddy terns excitedly splashed, dipping for and seldom missing the little rosy-red fish. From the helmet's coign of vantage, the commonest fish was the yellow-tailed *Xesurus,* moving slowly along in huge flocks. Large blennies are most amusing, scurrying over the rocks like field mice.

April 11th. At one o'clock this morning Mr. Tams, second officer, woke us. In less than four seconds everyone who had heard him was on the bridge. In the western sky was a rosy pulsation, now flaring high, now dying to dimness. At first we thought it must be on James, where the most recent volcanic activity has been, but it was eventually located on Albemarle.

For the last two days the wind has veered to the south and even to the southwest, blowing directly through the gap in the cliffs that is the mouth of the bay. This is an unusual quarter for the wind to come from in the Galápagos, almost unheard of, and may be connected with the volcanic outbreak. The anchor dragged and this morning we were not much more than a ship's length from the cliffs. If the southwest wind should strengthen, our position might be dangerous, and the heavy surf breaking so near made us uneasy. So the small boats were left on the beach, a quicker means of disposal than waiting to hoist them all aboard, and we left Darwin Bay, volcano-bound.

Five miles off the bay we made Station 38, with splendid results. Depth 550 fathoms. Two Petersen trawls brought up four-inch Myctophids, and an enormous Leptocephalus, as well as deep-sea Medusæ, shrimps, and some new Heteropods. A great many luminous Myctophids taken in a surface net after dark, and several of them lived for four or five hours.

We steamed at half speed all night toward the increasing brilliance of the volcano, which gradually assumed form. Very little sleeping done. Toward morning we could see through the glasses the actual molten spots. Bright moonlight, a tranquil sea, and a heavy low cloud reflecting the red-hot lava beneath it.

Noon position: Lat. 0° 18′ N: Long. 90° 03′ W.

April 12th. At dawn we were about six miles off-shore, and as the sun rose every vestige of color seemed to be wiped out, though the whole slope between the two northern mountains steamed and smoked. Three miles from land the soundings gave us a mile and a quarter, and two miles closer in, the water was over half a mile deep. Circled slowly all day as near the coast as possible, while a small boat took two of us ashore to try to reach the largest crater. Hours of inconceivably ghastly crawling, climbing, and falling over endless miles of crumbling, sharp-edged lava, under the equatorial sun, with heat pouring over us from above and below, brought us to one of the smallest of the fumaroles, where we encountered an invisible, almost odorless gas; before we could stagger into clearer air, we were almost overcome and deathly sick, and dared not attempt any further penetration, since we lacked gas-masks. The journey back to the shore was a nightmare, and in the midst of tumbling down ravines and creeping out again over rocks like heaps of broken glass, we were attacked by severe cramps in legs and feet, and made the last few yards practically on all-fours.

All night we lay off-shore and watched the glorious sight. As darkness came, the lava glowed deeper and deeper red, and from the crest of the ridge between the two mountains a long tongue of flame now and then shot up quivering against the sky. The whole long slope was dotted and smeared with fiery spots, and the huge mass of cloud that clung over the place reflected the furnace below it. So far as we know, no other eruption has ever been recorded from this northern end of Albemarle.

Noon position: Lat. 0° 04′ S: Long. 91° 11′ W.

April 13th. No anchorage here, so again went slowly up and down the coast all day, and towed trawls and nets, while a party went ashore to get moving-pictures of the little fumaroles that could be reached. At the tiny cove where landing was made, was a tide pool about ten yards across, where two moray eels were caught on hook with crab for bait. One (*Murœna clepsydra*) is a new record for the Galápagos. The small black yellow-tailed angelfish (*Pomacentrus arcifrons*) were abundant, up to six-inch ones, and two schools of fifty, and two hundred, *Querimana harengus* were also in this little tidal pond. The latter in defense formation were browsing on a rock, as close to each other as possible, gradually working down the chosen stone from the top, as a swarm of locusts might clear a field of grass. Red crabs were everywhere, also the smaller black-and-white spotted ones, and were quite fearless, sidling up and seizing the bait that lay on the rocks beside us. Gobies covered every submerged piece of lava, and we saw five or six *Eupomacentrus beebei* and a small *Holocanthus passer*. Several families of sea-lions, Galápagos gulls, pelicans and shearwaters were about.

To-night at dusk the molten lava was creeping down the slope in true Pompeiian fashion, and just over the shoulder of the mountain there must have been the greatest display, judging by the intense glow which was exasperatingly all that we could see. The bivouac fires of a tremendous army seemed to be scattered over ten or twelve miles of country, and as the hours passed the whole black incline became daubed with slowly-writhing scarlet streams creeping toward the sea. The sun set almost directly behind the ridge, and the changing of scarlet sunset into the rose and scarlet of cloud and lava was marvellous. It seemed as though a part of the sunset might have become entangled on these ravaged shores, while the rest went on its way over the rim of the earth.

The problem of the current on the east coast of Albemarle is a puzzling one. On the official maps a steady northwest stream of from one to two and three-quarters knots is given,—cold, Humboldt waters. This time the temperature has been no lower than elsewhere, and the organisms have not appreciably taken on a cold current character. The Captain discovered (and we going ashore in small boats verified) the fact that at the surface at least there is a tidal current. On the lowering tide the current sets strongly north, and on the rising tide it turns and sets as strongly southward along the coast, and at least ten miles out.

We steamed slowly northward after dark and at 8 p. m. put out surface nets, getting results quite different from those a few miles south. The Ianthina, Porpita and Glaucus brought to mind the Current Rip.

We seem to be establishing a record in equator crossings, as the Line runs through Cape Marshall and in our volcano observations we have gone back and forth till everyone has lost count of the times.

Raining to-night. We are now imputing the unusual season to the eruption.

April 14th. The current carried us swiftly northward and at midnight we shut off the engines and drifted. At 4 a. m. a half-metre surface net was towed for half an hour; the dominant organism was the deep-blue Copepod. The fish were young halfbeaks, two Coryphæna-like larvæ, two *Myctophum coccoi,* and twelve very small white larval fish, with large, semi-telescope eyes and spoon-shaped jaws. These early mornings the sea is a mirror, with low, oily swells that are barely perceptible, an almost colorless setting for the bright jewel of the crater.

Off Redondo Rock we put down a bottom dredge in almost two miles and got an astonishing collection. The huge net swung aboard bulging with an enormous load of lava, clay, huge crimson living corals, orange and pink starfish, scarlet shrimps, glass sponges, Hydroids, Crinoids, and about sixty huge sea-cucumbers, as icy cold from the chill of their native surroundings as though they were of the vegetable variety and had just been taken from the refrigerator. They were pink, purple, green, yellow and white, some smooth, others with long stems and bristles, or shaped like Turkish slippers.

At night the volcano appeared beautifully symmetrical,—a red cone tapering to a straight column of fire that joined a flat red cloud. During the day we heard low rumbles, but it may have been thunder instead of subterranean convulsions, as the lightning is brilliant to-night.

We are now steaming toward Abingdon.

Noon position: Lat. 0° 24′ N: Long. 91° 19′ W.

April 15th. Off west coast of Abingdon at 8:45 a. m. Most of the island a sheer cliff, partly covered with low vegetation, tapering off to each side,—to the north into a long-drawn-out dead black lava promontory. Deep water close to the cliff. One mile off sounded in 431 fathoms. Lowered bottom dredge and tangle but after towing for twenty minutes lost everything on the volcanic bottom. The cable was chafed white for 200 feet from shackle. Later put out a Petersen and half of it was torn away on a submarine peak. Fishing party in a small boat lost six spoons and most of their tackle to huge fish,—sharks and groupers, probably. They caught a twenty-one-pound black grouper and an eighteen-pound *Seriola dorsalis*. In the stomach of the latter was a scombroid fish, which had eaten a Zoea and two shrimps. A school of these large Seriola were jumping all round the boat.

The island is green and looks like an interesting place for study. A flock of sixty shearwaters were resting on the water off-shore.

Another Petersen put over with cheesecloth bag in bottom, and got some curious larval *Munidopsis,* a pear-eyed larval fish, and a medium-stalked *Stylophthalmus.* At 9 p. m. we tried making a deep haul without the deck-hands, and lowered a Petersen and a silk net together to 200 fathoms. Many interesting small things but no Myctophids nor Cyclothones. The best specimen a three-inch *Argyropelecus,* apparently a new species, close to *affinis* with mouth and eyes at an angle of 90° up.

Volcano still visible after dark.

Steaming slowly back to Tower, not wishing to get there before daylight.

Noon position: Lat. 0° 36′ N: Long. 90° 47′ W.

April 16th. Entered Darwin Bay at 7 a. m., with the smoke of the volcano faintly visible to the west. Many sharks and devilfish at entrance to bay as usual, and the rigging lined with brown boobies. Shore parties during day and diving in the shallow water directly east of the anchorage. The sharks show no interest in our presence, other than a mild curiosity.

April 17th. Most of us to the Crater Lake this morning. The way rises gradually over the usual rough lava with ordinary vegetation, where nesting boobies and frigates hiss at intruders. One owl flew close to our heads. We came on the crater as suddenly as though it had been a well, and found it about a half mile across, with the lake at the bottom of cliffs that taper down in successive slips. Looking down at the water, the centre of the lake is clear olive-green, the shallow part at the rim sage green, and scattered irregularly along about half the shore-line are dense blue-green mangroves. From the northern rim, the sea on three sides of the island was visible, and from the west the *Arcturus* lying in the bay could be seen. We had a hard time getting down, burdened with nets, bags and buckets. Close to shore the mud was only shoe-deep, but beyond the mangroves we sank in to the knees, stirring up an overpowering smell of rotting matter, animal and vegetable. Green algæ thickened the water in many places, and everywhere were untold myriads of small water-striders, and aquatic Hemiptera, the "water-boatmen." Small round bivalves covered every strand of the weed. We found a spot where deep water was closest to shore and swam part way across the lake. The water was intensely salt, so that we floated high in it, and it made the eyes smart. Took a salinity bottle full for sample (Fig. 67).

In the crater was an almost pure culture of red-footed boobies nesting in the mangroves, in nearly every case a bird in the white phase mated with a brown one. There were at least six birds that were brown with white scapulars and inner flight-feathers, two of which were sitting on eggs. Two pairs of yellow warblers were singing, and a flock of about twenty-five turnstones were feeding on the insects. A few frigatebirds were nesting high up on the crater-sides.

Along shore was a long-stemmed, jointed, floating grass, to which were fastened millions of eggs and developing water-striders.

Diving in the afternoon.

April 18th. Long diving sessions most of day. Took down crab bait and thousands of fishes ate from the hand. Literally clouds of small fish, such as *Paranthias, Pomacentrus,* and *Thallasoma,* swarm about twitching off bits of the bait; the sensation is of an aquatic version of feeding pigeons at St. Mark's. Also some successful harpooning of fish done under water, a large blenny, four yellow-tailed *Xesurus,* a beautiful specimen of *Chœtodon nigrirostris,* and the blue-striped golden *Evoplites viridis* being obtained in this way. The larger fish do not seem so hungry, or perhaps it is that they are more cautious, but the small ones are quite fearless and brush against the diver's arms and legs as unconcernedly as though he were a familiar sight.

Saw *Evoplites* two and a half feet long. They feed on crabs, taking in whole sections of leg-joints.

April 19th. Shore parties as usual, fishing parties in bay, and a trial at diving in one of the deep rock pools to the west of our landing beach. Very different in absolutely quiet water from the surge along shore that scrapes one along helplessly.

A friendly penguin was added to the passenger list.

April 20th. More diving and fishing from the rocks under the cliffs. One method of collecting is to have a bucket lowered to the diver after he reaches the bottom; this he fills with rocks, which are placed in a large tub on deck of ship, and in a day or two all the creatures have crawled out of their nooks and are easily obtained. Mollusks, worms, annelids, squillæ, isopods, and many species of crabs have been collected in this way.

At 4:30 p. m. we started for Hood Island. It is like leaving home to go from this place that is so familiar to us,—birds, cliffs, sandy beach, and now even the underwater portion of it is not entirely unknown. Large rays were leaping in the bay as we steamed through the narrow opening, and the sea side of the eastern point was solid white over a large area with white boobies.

At 9:30, south of Tower, towed a surface net for twenty minutes and caught 1288 Myctophids, of which 88 were *M. coccoi,* and the rest *M. affinis.*

April 21st. Awoke to a heavy swell and a squally, rainy day, most unlike this region. Passed Barrington early, and Chatham in the afternoon, dim and hazy in rain-clouds. Sounded at 3:30 p. m. in 173 fathoms and a Petersen trawl yielded nothing but a few very remarkable crab larvæ, some with enormously lengthened fore and aft bars, and others with radiating rods with strange swellings on some of them.

Several albatrosses flew past the ship and we could see the eastern end of Hood, where they nest. Occasional large flyingfish, and black shearwaters.

Five miles south of Hood sounded in 401 fathoms and after dark sent down the small Petersen and a metre net to 200 fathoms. Got a huge transparent Isopod, *Cystosoma,* alive and perfect. In a half-metre net towed at the surface at the same time, we took five hundred Myctophids, mostly *M. affinis.*

Earlier in the afternoon a metre net caught a beautiful mass of small colored medusæ, stalk-eyed shrimps, and radiating, highly colored radiolarians.

John Tee-Van was sitting in front of the big aquarium this evening when the glass broke and deluged him with fish and water.

Noon position: Lat. 1° 00′ S: Long. 89° 41′ W.

April 22nd. A grey day with the *Arcturus* doing some reminiscent rolls. Now and then we clutch at things and brace our feet in almost Atlantic style. Thirty-three miles south of Hood sounded in 1820 fathoms, then ran a line of thermometers to 3000 metres, where we found 36.5° F. The instruments were icy and the water-bottles frosted.

Put out a Petersen and a metre net at 1200 fathoms, an otter trawl and a metre net at 800 fathoms and a metre net at 400, taking two hours. We wallowed badly, but after towing for two hours, and consuming two hours to bring in the cable again, found everything in good shape except the otter trawl, which was all wound up on the main cable. The haul as a whole was excellent and there were many new fish.

Noon position: Lat. 2° 00′ S: Long. 89° 37′ W.

April 23rd. There being no signs of the Humboldt Current yesterday, we went slow on both engines all night to the south, and sounded this morning at 5:30 in 1835 fathoms. Then we took six temperatures down to 500 metres.

The drift to the east last night was almost nil, as compared with that the preceding night, which was strong to the northwest. This shows either that the cold Humboldt Current has been driven far to the southward, or it is temporarily overlaid by the Panama Equatorial Current.

Swell continues, sky is overcast, with at least three rain squalls in sight to the southward. A shearwater flew on board at 4 a. m.

At 9 a. m. sent down Petersen and metre net to 500 metres, a second metre net to 300, and a third to 100 metres. In the second and third nets were masses of the pale salmon-pink shrimp and copepod plankton which we get at the surface every evening, but no Myctophids. *Cyclothone,* and *Vinciguerria* were the dominant fish, and there were also many larval forms of several unidentified species.

In afternoon put half-metre net on end of a Petersen and sent down to 800 fathoms. Complete failure, only a little plankton and one black *Cyclothone* resulting.

A petrel flew aboard at 8 p. m.

Noon position; Lat. 2° 33′ S: Long. 89° 44′ W.

Three or four tows in evening at surface. First brought in twenty Myctophids, but at 9 p. m. there were none in net. A dozen tiny, elongated fish larvæ were taken, one of which had been swallowed by a Sagitta. *Holocentrus* larvæ taken, and some *Leptocephalus,* three of which were enormous.

April 24th. All day spent putting down, towing and bringing up bottom dredge, and at 4 p. m. when it reached the deck, it was opened and its contents studied.

Noon position: Lat. 1° 44′ S: Long. 89° 39′ W.

April 25th. Bottom dredge down this morning in two miles of water, and recovered it before noon. It contained a vast heap of sea-cucumbers, icy-cold from the depths, and not much else. Steamed north for Hood as soon as the dredge came in. Found a tremendous surf beating on southeast side, and went round to northwest, where all swells and rough water ceased. Anchored in twenty-one fathoms in Gardner Bay, one of the loveliest spots in the Galápagos. A long sand-beach was a favorite sunning place for sea-lions, and hundreds of doves fluttered along shore. Saw black hawk, many mockingbirds, and a few finches.

Small boats were quickly put over, and everyone scattered to explore, collect and fish. In the evening the spot-light beside the lowered gangway attracted large, gorgeously colored flyingfish by hundreds, whizzing out of the water like bullets, striking with a crash against the side of the ship, falling on the main deck, and filling the boats that were moored to the boom. The bottom of the gangway was really quite a perilous post, as a fifteen-inch fish going full tilt through the air is a missile not to be despised. There were also half-beaks and pipefish, and many smaller species, which we scooped up with nets, and found to be most interesting, but difficult of identification. Sea-lions were at the edge of the circle of light, occasionally rushing into view to seize a fish from the numbers milling about the ladder.

April 26th. Diving-helmet in action again, in the lee of Gardner island. Found water slightly colder here. Brought up many rocks from the bottom, for their covering of invertebrates, and later found Balanoglossus and a host of worms and crabs. In one spot, small, pale green sea-urchins were dotted everywhere, so that it was not possible to sit on any stone, and in another place plucked six large holothurians.   Three or four sharks came around to look at us, but except for a baby, about two feet long, that played with the end of the harpoon, none of them showed any disposition to molest us.   Sea-lions once or twice gave us a start, as they shot down from the surface to look more closely at us.

On Gardner there were finches singing, doves were abundant, and mockingbirds as tame as usual.   Saw some of the Tropidurus lizards, which seem larger here than elsewhere in the group, and with more yellow in the markings, such as a line down the back.   Three Amblyrhynchus caught are much more reddish than any we have seen on other islands.

Scattered about were small greenish rain pools deep among the rocks, and in them were countless Branchipus and big ostracods. On Hood Island much fresh water was found, one pond a half-mile in length being seen.   Fresh-water crustaceans were collected, including Apus.   Insects were numerous, and Serge got a good many grubs and caterpillars.   There were many goat skeletons, and one of the sailors shot a large male.

Attempted to use the small dredge from one of the motor-launches, but found the bottom of the bay too irregular to make it possible, as it is heaped with lava almost as thickly as is the land.

This evening flyingfish were again with us in flocks, and dozens of the green Calosoma beetles flew across the mile of water to the ship.   Very few moths came.   A twenty-minute plankton haul from a motorboat brought in a greyish pink mass of copepods, megalops, ctenophores, siphonophores, doliolum, sagittæ, pteropods, lucifers, schizopods, radiolarians and stalk-eyed shrimps,—quite a pelagic haul. This emphasizes the excellence of this place as a site for a year's intensive study.

April 27th.   Exploring party to small, nameless island in bay, which is now named Osborn Island.

April 28th.   Another party to albatross rookery brought back two adult birds alive for the Zoological Park.   Another small island explored.

As coal and water are getting low, we start toward Panama tonight.

April 29th.   Took sounding and temperatures this morning.   The sounding wire broke and lost over a thousand feet of wire.   Petersen sent down to 800 fathoms, a metre net to 400 fathoms.   In the Petersen were large maroon Medusæ, with very long tentacles, huge scarlet shrimps, and some medium-sized ones, a small *Melamphaes nigrofulous,* and some pink sagittæ.

The metre net held *Oneirodes, Melamphaes,* another mucous-headed fish, and small shrimps, a few worms, a very large scarlet ctenophore,

FIG. 67.—LAKE ARCTURUS.

An intensely salt crater lake discovered by Dwight Franklin in the center of Tower Island, Galápagos.

Fig. 68.—The *Arcturus* Returning from her Six Months' Voyage, Flying her One Hundred and Eighty Foot Homeward Bound Pennant.

isopods, copepods and sagittæ. The dominant color of all the invertebrates was red.

Many flyingfish to-day, but none of the huge ones that were so plentiful in Gardner Bay.

We are now looking for the Current Rip.

Noon position: Lat. 0° 02′ N: Long. 88° 23′ W.

April 30th. A placid sea, and a very warm day. No Current Rip to be found so far, and as we have twice crossed its former path it has apparently passed out of existence.

Noon position: Lat. 2° 47′ N: Long. 87° 16′ W.

May 1st. Our one remaining ice-box seems to be dying, so that we drink tepid water, and several hundred pounds of meat are in peril. The Rip is lost, and the Captain thinks the wisest plan is to run straight for Panama for repairs to ice-plant, lights, launches and all.

Two wonderful hauls to-day, in which we obtained for the first time the *Gasteropelecus* which adorns our house-flag, as well as the largest example we have ever caught of *Chauliodus,* about ten inches long.

Noon position: Lat. 3° 57′ N: Long. 86° 48′ W.

May 2nd. Sea like a sheet of glass, and we are going full speed for Cape Mala. Two big turtles, and an olive-footed booby sitting on a small log passed at 11 a. m. Under the log hundreds of fish were swimming. During the day a large petrel passed close to the ship, swimming quite fearlessly. Saw three sea-snakes, and many logs with attendant schools of fish, and caught a small Coryphaena from the boom-walk. As it was pulled in, several larger ones followed it hungrily. A four-foot *Pyrosoma* and a shark were the only other creatures seen, except flyingfish.

At 1 p. m. sounded in 1690 fathoms, and made a successful haul.

Noon position: Lat. 4° 52′ N: Long. 84° 42′ W.

May 3rd. An incredibly smooth sea, with abundance of life,—flyingfish, great schools of tunny and dolphins, many white boobies and shearwaters, extensive patches of dark brown, sponge-like algæ, porpita, and so on. At 10 a. m. an eastern cliff-swallow came aboard, and for several hours a large-billed water-thrush was on the ship. (Location 5° 47′ N., 82° 58′ W.)

The albatrosses are feeding from our hands as though they had always eaten that way, and allow us to pet and stroke them. They stand up most of the time, drink nothing, but enjoy a thorough spraying every morning. Now and then they go through a portion of the dance, clattering their beaks, or bowing, or raising their heads straight up.

A half-hour metre net at the surface at 8 p. m. yielded blue Copepods, and scanty grey-pink plankton, many small Porpita and an amazing number of small fish. Two large Myctophids, several small ones, a wonderful copper-and-silver round fish with enormously elongated spines, and a tiny creature like a swordfish with two pale-blue spots in the back, besides brilliant flyingfish of several species, a half-beak with pectorals as long as a flyingfish, many small Coryphænæ, and so on. Also six tiny squids, and sagittæ, Halobates, Ianthina.

Noon position: Lat. 5° 49′ N: Long. 82° 46′ W.

May 4th. Crossing Panama Bay all day. Busily cleaning laboratory in preparation for many visitors expected when we dock. The place looks unholily neat.

Noon position: Lat. 7° 13′ N: Long. 80° 20′ W.

May 5th. Reached Balboa at 8 a. m. and tied up to Pier 16. Spent five days preparing for another six weeks in the Pacific. Find from our letters and newspaper clippings that we were lost to the world for ten days, but as we did not know it, we were not worried. We did realize that we could not pick up any wireless station, nor even relay through another ship, but did not think anyone would be excited about it.

May 10th. Left Balboa wharf about 2 a. m., and anchored down the bay. Two essential firemen had deserted, so we removed temptation from the rest by lying well away from shore while the agent rounded up substitutes. These were dragged aboard at 6:30 in the evening, and we started for Cocos Island. Four submarines steamed past us, homeward bound under a golden sky.

May 11th. A grey day, passing the Panama coast, with its jungle-clad mountains. Tide rips and heavy swells about Cape Mariato. Water rich in life, great schools of mackerel leaping, and feeding on smaller fish, while hundreds of shearwaters circle low and swiftly just above the water. Caught a twenty-two pound tunny on spoon from boom.

Dragged a meter net at 8 p. m. and got about twenty Myctophids, many small jellies, numerous *Phyllosoma,* and a fair amount of pale pink plankton. A pail in the end of a small otter trawl caught nothing. The fish in meter net were 7 *Myctophum laternatum,* 11 *Myctophum affinis,* 3 *Myctophum coccoi,* 2 *Myctophum humboldti,* 5 larval fish, 2 large *Leptocephalus* (135 and 200 mm. long) and 2 small flounders.

Noon position: Lat. 7° 10′ N: Long. 80° 18′ W.

May 12th. Calm as a mirror, with a circle of blue-black rain-clouds around the horizon before sunrise and lines of rain showing here and there.

In evening attempted to make four fifteen-minute hauls with meter net from 6 to 7 p. m., but had to stop after second one because of the mass of Ctenophores (*Mnemiopsis*) which filled the nets with gallons of almost solid jelly. Only two or three very small fish in the mass. At this time we passed a small current rip running north and south, and perhaps the abundance of jellies was connected with this.

At 8:30 p. m. ran out otter trawl with half-meter net attached and secured about the same forms as in metre nets at 8 and 9 o'clock. A hundred or more Myctophids, almost all two inches or more in length, and the majority *Lampanyctus macropterum,* a form new to the Pacific, seven giant *Leptocephalus,* and two wonderful Scopeloids, with large teeth, long barbels and the dark brown skin covered with small light-organs.

Noon position: Lat. 6° 33′ N: Long. 82° 43′ W.

May 13th. Station 66. Successful hauls with a Petersen at 600, and three metre nets at 600, 500, and 300 fathoms. From the Petersen we obtained:

1 *Stomias colubrinus* (145 mm.), 4 *Melamphaes mizolepis,* 1 *Melamphaes nigrescens,* and 2 *Lampanyctus macropterum.*

The metre net at the same depth yielded *Lampanyctus oculeum,* and the 300 fathom net brought in several *Vinciguerria lucetia.*

Worked hard on yesterday's material, and in evening surface haul (8 p. m.) got almost no fish, only 2 *Myctophum coccoi,* I large Leptocophalus and 2 *Lampanyctus macropterum.*

Noon position: Lat. 6° 25′ N: Long. 85° 06′ W.

May 14th. The following is a list of fish caught in a typical Pacific haul, the list belonging properly to future technical papers. Soundings and temperatures at daylight. Half-hour haul at 5 a. m. yielded large numbers of Halobates and pale pink plankton. The fish were:

4 *Myctophum coccoi.*
1 halfbeak.
2 young Coryphæna.
1 small Coryphæna-like, dark-finned fish.
1 yellow-banded flyingfish.
1 oval four-banded fish.

From 9 to 11 a. m. towed a Petersen and four metre nets, with following results:

Metre net at 300 fathoms.
   Numerous Vinciguerria and small white Cyclothones. These resemble each other and form a semi-transparent white zone of fish.
Metre net at 400 fathoms.
   400 white Cyclothones.
   200 black Cyclothones.
   Numerous small *Myctophum laternatum.*
   1 very elongate long-barbeled fish.
   Large white transparent Octopus, with two circles of light-organs around eyes.
Metre net at 500 fathoms.
   Net torn, contained only one white Cyclothone.
Metre net at 600 fathoms.
   1 seven-inch *Chauliodus sloanei.*

1 large silver eel.

1 *Nemichthys.*

1 *Melamphaes.*

1 *Stylophthalmus.*

1 *Melamphaes mizolepis.*

Petersen at 600 fathoms.

1 small white fish, with enormously elongated thread pectorals.

1 long-jawed *Leptocephalus.*

2 *Melamphaes nigrofulvus,* one with stomach distended by huge fish.

1 *Idiacanthus antrostomus.*

1 65 mm. *Chauliodus sloanei.*

1 *Melamphaes mizolepis.*

1 *Melamphaes megalops.*

1 70 mm. *Stomias.*

300 black Cyclothones.

25 white Clyclothones.

Cocos sighted before noon but we are roaming around in the vicinity to-night and will seek anchorage at dawn. "A bold coast," says the Captain and prefers daylight by which to verify what the charts say about Chatham Bay.

Dozens of small tunny escorted us all day but refused to take a hook.

Noon position: Lat. 6° 15′ N: Long. 86° 46′ W.

May 15th. A wonderful sight at seven this morning. Hundreds of porpoises leaping around the ship, boobies flapping in the rigging, a rainbow arching before us, where Cocos lay sombrely under heavy rainclouds, with snow-white terns silhouetted against its gloomy shores. Mad attempts to harpoon a porpoise resulted only in much exercise for everyone, including the porpoise.

Anchored in Chatham Bay and everyone seized handfuls of biscuits and rushed ashore, ignoring luncheon. Found the island a mass of tangled vegetation, the only clear space being the narrow strip of beach. Even the hill nearest the sea, that from the deck seemed to be covered by a smooth lawn, proved to be overgrown with tall sharp-edged grass, in which it would have been easy to lose oneself except for the slope of the ground sea-ward. So high and tough and closely-matted is this growth that the easiest way to progress is by falling through it, literally pitching forward and so pressing down a sort of trail, along which it is possible to flounder. The feeling is that of wading through a gigantic haymow.

The crew were swimming in the bay; the dynamiting party acquired new species of fish from the reefs by explosive methods, while others investigated the shore and river, seeking for insects, birds and water-life.

The only signs of man were the names carved on the boulders in the river-mouth, and on top of the hill a few rotted boards, pieces of corrugated iron, and ancient hand-cuffs.

Everyone on board for dinner, much lacerated as to arms and legs.

May 16th. Another busy day for everyone, diving, fishing, collecting and exploring.

May 17th. All-day rain. Much diving done in the morning, a rather dismal proceeding when you emerge into a drizzle and sit damply in already soaked bathing-suit. The wonderful coral reefs were reward enough, however. At one spot they were all in the shapes of Gargantuan mushrooms, with bright fishes floating in and out around them.

In the evening we were invaded by hundreds of boobies, who seemed to find the lighted ship more attractive than the wetness of the island. They crashed aboard, rushed into laboratory and cabins, screaming and flapping. It was a chaotic scene, as we dodged the broad flailing wings and sharp beaks, and threw them, hissing like angry geese, over the rail, only to have them return a few seconds later. The small boats, tied to the boom, were full of them, so was the rigging, and they squawked and blundered all over the decks, ejecting large fish from their crops in their excitement. Then a heavy squall came up, with sheets of rain, dashing spray, loose boats and frantic birds,—the wind blowing half a gale and the sailors shouting as they hoisted the boats on deck and made everything fast. It was a turbulent night.

May 19th. At daylight found that the *Albatross* launch was gone; she had been anchored in the bay as a convenience for diving, and must have chafed through her moorings and drifted away. One of our two diving-helmets was on board. The first officer went off in the *Pawnee,* the other launch, to search for her, and one squall after another kept us worrying about him until he finally returned late in the afternoon minus the *Albatross.*

May 20th. Not a drop of rain all day, wonderful to relate. Half a dozen dynamite charges have proved the most productive methods of getting new species of fish at this place and there follows a more or less typical list of one day's results:

   10 *Dermatolepis,* 1½ to 2 feet long.
    5 *Caranx melampygus.*
    2 Barred-fin Surgeonfish.
    1 White-tailed Surgeonfish.
    2 Puffers (one yellow, the other black-and-white).
   12 *Evoplites viridis.*
  104 *Pomacentrus arcifrons.*
   14 *Paranthias furcifer.*
    4 *Pachygnathus capistratus.*
    8 *Lutianus jordani.*
    4 *Holocentrus suborbitalis.*
   13 large *Myripristis murdjan.*
   23 small *Myripristis murdjan.*
    1 *Zanclus canescens.*
    1 *Chætodon Nigrirostris.*
    1 Ostracion sp.

May 21st. Another day of fishing and diving, with tubs of fish to identify and preserve as a result. Exploring party to Wafer Bay, where are the remnants of the settlement that August Gissler had here. Attempts made to harpoon giant rays from the *Pawnee* launch, but the rays always vanish by the time the launch and gear go in pursuit.

May 22nd. Everyone ashore to collect at low tide. Got several tiny morays and two species of *Thallasoma*, as well as parrotfishes, Moorish Idols, puffers and a small trunkfish. Crabs are abundant and varied, but almost no sea-anemones and few sponges. Shrimps common and many curious clicking shrimps. More diving and more dynamited fish. In the evening one of the deck-hands harpooned two sharks from a small boat tied to the boom. One was *Carcharias platyrhynchus*, about six feet long, the other a four-foot *C. galapagensis*. The first had eaten one *Lutianus jordani* and two *Paranthias furcifer;* the second one *Paranthias* and two *Pomacentrus,*—but they were all dead fish which we had thrown overboard.

May 23rd. Cloudy, squally day, water very rough. Collecting and seining on beach and in river. Got about three pailsful of sandfish at the mouth of the river, and up the stream captured over a hundred gobies of several species. There were very few of these latter in the quiet side pools, but many in the swift-rushing main channel. Besides these there was little life to be found, except small shrimps, which were common, and an occasional large-clawed blue crayfish.

We have seen several rats and caught one, very mangy and thin. One wild cat seen on beach, eating dead fish, which was also the food of the rats. Three or four large pigs have been shot at by the sailors and once we saw a very small one.

Tremendous rain from four p. m. all the rest of day.

May 24th. A dank and dismal all-day rain. In the morning managed to poison some of the tiny tide-pools with copper sulphate and so obtained a number of small fishes of kinds that we have vainly tried to take with nets. About 2 p. m. got up anchor and steamed round north and west sides of island. There are dozens of cascades pouring into the sea over high cliffs, bursting from the cover of thick woods, and leaping down smooth rocky slides. By the time we were off Dampier Head the rain, fog and clouds were so heavy that we could not see much. Stood off to southward.

May 25th. Establishment of Station 74, at 4° 50′ N. Lat.: 87° W. Long. A hot sun, welcome change after ten days at Cocos. A splendid haul in morning with five silk nets and a Petersen, repeated in afternoon.

Everyone tired from strenuous days on Cocos, and with bruises and cuts and sprains to cure.

May 26th. Vertical hauls all day to-day, to establish controls on zones of life. Enough sea to roll the ship about when lying idle.

May 28th. An otter trawl put down this morning to the bottom. It came up wound round the boards at its mouth, but in spite of that it contained the first specimen we have captured of the Macrurids. It seems strange that we have not had these before, as other expeditions appear to have taken them very often.

A deluge of rain from moon to dark.

May 29th. Two otter trawls on bottom to-day, the first very good, the second much better, containing huge black eels, *Bathypteröis* in perfect preservation, Macrurids, bat-fishes and the more common deep-sea fish, as well as sponges, shrimps and all sorts of invertebrates.

At 5 p. m. black clouds and a gale of wind preceded a downpour which lasted all the evening.

May 30th. Two good hauls with otter trawls on bottom to-day, bringing in batfishes, eels, Macrurids, Brotulids, and starfishes, sponges and shrimps by the hundred.

May 31st. Fair most of the day, showers at night. Lost an otter trawl trying to put it over without the crew, as this is their day of rest but the scientist knows none. Then put down the coarse bottom dredge, and at 2 p. m. started a twenty-four-hour series of plankton hauls. Put out a surface net every half-hour and left it out thirty minutes.

June 1st. Torrents of rain all day, clearing about 5 p. m. Enormous amount of specimens and data acquired from twenty-four-hour plankton hauls. The bottom dredge brought in a fairly good haul.

June 2nd. A rather disastrous day on the whole. The 80-foot dredge was put over and lost, various nets went wrong, and, two Petersen trawls brought in a sum total of one Salpa. Squalls of wind and rain and a terrific downpour in evening. At dinner-time a large school of blackfish gathered round the ship, so near that we had a good chance to see them as they came up to blow.

June 3rd. A hurricane blew a deluge into the cabins on the windward side all night. Everyone busy alternating between shutting the doors and gasping for air, and opening them and drowning. A Petersen and five metre nets down in the morning, and a dredge in the afternoon, with good results. As soon as the dredge was in, we started for Albemarle and Tagus Cove. All hoping for some sunshine to-morrow.

June 4th. Grey days and rain-squalls still the rule. Plodding along all day without hauls. At 9 p. m. a sounding showed that we had gone off the edge of the Cocos plateau very abruptly.

Noon position: Lat. 3° 53′ N: Long. 87° 13′ W.

June 5th. Still cloudy and grey, though no rain. No hauls to-day. Thousands of purple tunnies leaping round the ship and refusing to be caught as usual. Many shearwaters about, banking and dipping their wings.

Noon position: Lat. 2° 39′ N: Long. 88° 31′ W.

June 6th. Raised Abingdon and Bindloe soon after lunch and passed the former by moonlight. Anxious eyes fixed volcano-wards as we approach Albemarle, but no sign of activity to be seen yet. Boobies are with us once more, and one frigatebird,—the first since Cocos.

Noon position: Lat. 1° 06′ N: Long. 90° W.

June 7th. Up at 5 a. m. to see the northern point of Albemarle, everyone, even the Captain, astonished that we reached there so soon. A current evidently picked us up in the night and fairly hurled us along, doubling our speed. Far down the central ridge between Mts. Whiton and Williams was a column of smoke, the second mountain veiled in smoke or heavy mist and two large columns of smoke ascending from it. Saw several huge sharks, one apparently helpless, lying on its side.

Dropped anchor in Tagus Cove just before lunch. Fish taking bait even before we had come to a standstill. Almost everyone ashore, or diving, or photographing, and a busy evening in the laboratory to follow.

June 8th. Diving in the morning, with penguins, cormorants, sea-lions, pelicans and great blue herons all round the boat, watching the performance with interest. Not a very good place for diving, as all the rocks are black, the water not very clear, and very deep. Enthusiastic anglers brought back dozens of huge groupers, and some barracutas and mackerel. Then the dynamite squad returned with tubs of fish of all sizes, and the deck became a fishy shambles, with the scientists identifying and preserving those of interest other than gastronomic, and cook and steward frantically cleaning the groupers and mackerel for ice-box preservation.

At the gangway after dark we caught *Sphyræna, Coryphæna, Menidia starksii* and *Hemiramphus saltator*.

June 9th. Up anchor at 8 a. m., and round the northern side of Narborough. The most desolate island of them all, it seems, with hardly any vegetation, and the lava cliffs of the shore-line rising to long black slopes that lead to the central crater, draped in clouds.

What we suppose must have been a whale-shark rose alongside the ship about noon, floated there for a few seconds and then sank again. It must have been almost forty feet long, as Dr. Gregory happened to be standing on the bulwark at a spot even with the end of its tail, and John Tee-Van was abreast of the head, so it was possible to measure the distance.

Trawling in the afternoon, with good results.

The volcano is faintly visible to-night, and all day a dim pink glow hangs over the spot, reflected in low cloud banks.

Noon position: Lat. 0° 17′ S: Long. 91° 34′ W.

June 10th. Diving in a wonderful place to-day, amidst sea-weed so tall and thick it was like a corn-field. The trawls brought in two *Oneirodes* alive, that gave an exhibition of lighting in the dark-room. Petersen and metre nets in the afternoon.

Some one should start tours to this part of the world and advertise, "Spend the summer on the Equator and keep cool." It is really chilly here.

June 11th. An 18-foot ray harpooned from a small boat this morning and it took two hours, three men, four harpoons and a shot-gun to land it. Frantic excitement, with boats milling about, shouts and shots and scurrying, before the creature was hoisted aboard the *Arcturus*. The deck now looks like a slaughter-house, as the victim weighed over a ton, and dissection on such a large scale is very messy. A thirty-pound embryo taken from it, in perfect condition.

Diving and dynamiting proceeding as usual. The following list is the result of two discharges:

    25 *Holocanthus passer.*
    12 *Xesurus laticlavius.*
    39 *Paranthias furcifer*—small.
     6 *Paranthias furcifer*—medium.
    24 Orange-pectoraled Pomacentrus.
     1 Brown Pomacentrus.
    15 *Apogon atradorsatus.*
     3 *Bodianus eclancheri* (black).
     1 *Bodianus eclancheri* (orange and black).
     2 *Bodianus diplotænia.*
     3 Mottled Groupers.
     1 *Anisotremus.*
     1 Red *Epinephelus.*

June 12th. No rain—but no sun. Twenty miles off Narborough, to the west. At 6 a. m. sounded in 1900 fathoms, and at 8 a. m. put over a Petersen at 1000 fathoms, and five metre nets at 800, 600, 500 and 400 fathoms. Excellent hauls, including seventeen new species of fish. Worked south of Narborough during afternoon, and to-morrow go to north of Albemarle.

Last night a stormy petrel flew on board,—the dark phase of *Oceanodroma leucorrhea,* with no white on the rump. Confused voices of seabirds crying through the darkness all night.

The last piece of ray went overboard to-day and decks scrubbed.

Noon position: Lat. 0° 47′ S: Long. 91° 41′ W.

June 13th. Captain feeling jocose this morning, blew the foghorn at dawn and pointed out a tropicbird! Sounded in 1720 fathoms, on the equator, Long. 92°. A current rip very strongly marked as far as the eye could see, coming out of the strait between Narborough and Albemarle from the southeast, and after reaching the northern point of Albemarle, curving around to the southwest. This has probably something to do with the smooth backwater or eddy north of Narborough, which we found so full of life. The water to the north of the loop was very green and rough, while that to the south was blue and smooth. It looked as if the latter were flowing rapidly northward and pressing against or flowing under the green north water, causing the whitecaps to break in a southern direction. Trawled through it and make good hauls (Station 87) but there was not the surface life to make it as interesting as our first Current Rip.

Steamed around the northern tip of Albemarle and south toward Cape Marshall, running into an immense school of tunnies which were jumping eight or ten feet in the air and setting up a line of foam like breakers, two or three miles long. Sometimes twenty would be leaping in one spot.

In the evening with no moon, a rough sea (after we passed the shelter of the island's lee) and with cloudy skies, we made four enormous hauls of *Myctophum coccoi*. There were hundreds of them in each net. The estimate of 1437 square feet covered by a half-metre net in its course is probably too great by half, as the net is half out of water all the time. The speed was slow on both engines.

Noon position: Lat. 0° 07′ S: Long. 91° 49′ W.

June 14th. At breakfast time white spouts sighted, apparently coming from the sea beyond Cape Marshall. Heavy head wind and current made our approach slow, but little by little we realized the origin of the puffing masses that rose from the water. In the two months since we were here before, the lava has been working down the slope until it reached the steep cliffs of the shore, and here before us were nine great cascades of molten rock gushing from the face of the black coast, and dropping straight into the sea. Immense columns of steam were blown by the strong wind across the land, so that the cataracts were not obscured and we could watch huge pieces of the cliffs crumble under the pressure from behind, crashing outward to release fresh torrents of red-hot lava, that spouted like water from a culvert. Now and then submarine explosions from too-rapidly-cooled lava threw great lumps of glowing rocks above the breakers, that hissed and turned to steam as they dashed against the scorching shore.

The sea was choppy with tossing whitecaps; along the coast the water was vivid light green, where it was heated by the lava; a line, distinct as though painted on a floor, marked the beginning of the deep blue, normally cool ocean. So sharp was the demarcation that when the *Arcturus* was within a quarter-mile of land, as near as we dared venture, and lying directly across this line, her bow was in the green water at a temperature of 99°, and her stern was in blue water which registered 78°. As molten lava reached 3000°, the ocean under the cliffs was literally boiling. A sea-lion flung itself in agony from the scalding immersion, five times leaping all clear, and then

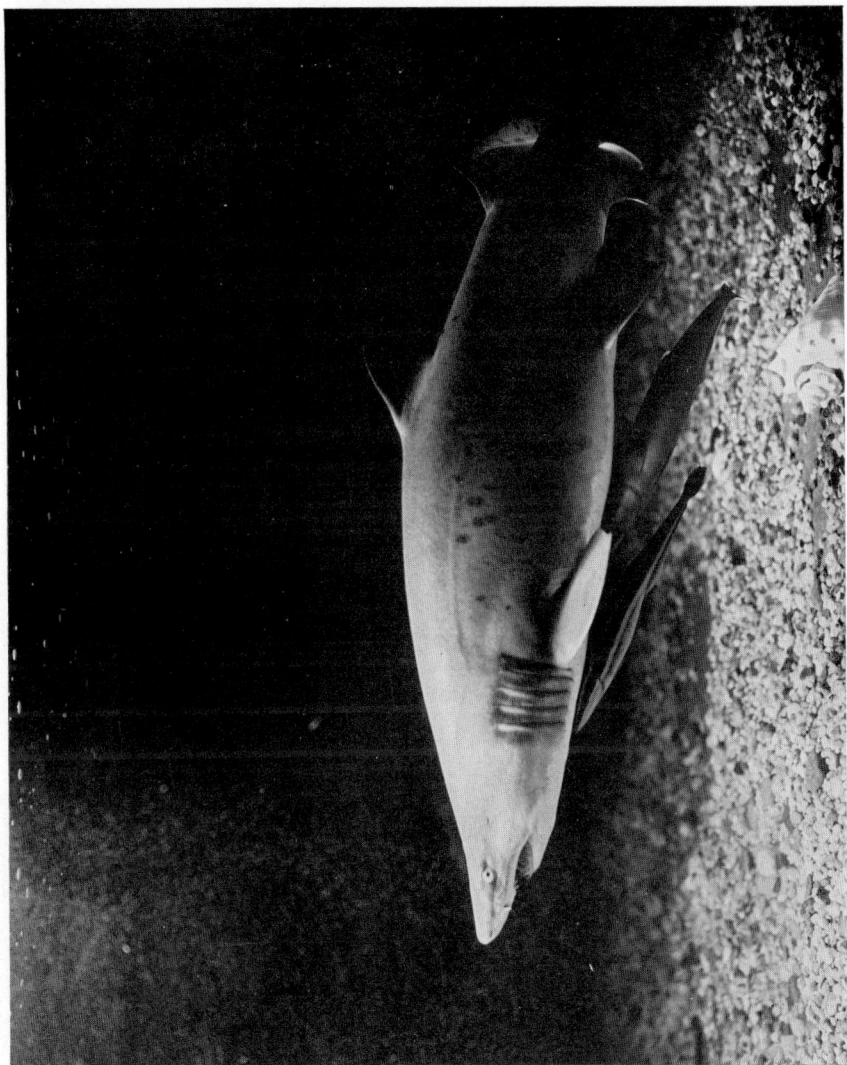

FIG. 69.—SAND SHARK WITH TWO SHARK SUCKERS CLINGING TO IT.

seen no more. Shearwaters and frigatebirds stooped through the vapor to snatch at fish floating in this gigantic cauldron, and we saw dead petrels and shearwaters that had ventured once too often to this tempting feast.

We circled in front of the spectacle all day, while binoculars, cameras and paintbrushes were busy. The sea was too rough for the launching of a small boat, so no attempt to land could be made. We feasted our eyes on the stupendous sight and made every possible record, reluctantly convinced by the captain that our impending water shortage made it imperative to start for Panama that night. About 8 p. m. we steamed away, watching the superb color as it receded, and longing for more time and further opportunity to study this marvellous display.

Noon position: Lat. 0° 03′ S: Long. 91° 12′ W.

June 15th. To-day the steering-gear broke without warning. It may be just as well that we left the volcano when we did, as the strong onshore wind and current might have driven us into one of those cataracts of molten rock,—and the *Arcturus* is a wooden ship. Steering with the wheel aft as soon as it could be put in commission. Condenser is out of order, so no stops for trawling to-day.

Passed in sight of James, Bindloe, Indefatigable, the Daphnes and Eden and at 3 p. m. we passed not far from Tower, to the southward, and sadly looked our last at the Galápagos.

Noon position: Lat. 0° 03′ S: Long. 91° 12′ W.

June 16th. A beautiful day, with some showers. A small school of real porpoises rolled along, going northwest and barely coming out to breathe.

Steering gear and condenser still out of order.

In the evening a white-faced petrel, *Pelagodroma marina,* flew on board, and others called plaintively in the distant darkness. Later one of the sailors caught a sooty tern.

This morning saw the last of the large maroon-winged flyingfish which are so common around the Galápagos and especially near Hood. In the afternoon only the small, clear-winged species seen.

Noon position: Lat. 1° 40′ N; Long. 88° 17′ W.

June 17th. Perfect day, slight swell. Tunnies in numbers acting as escorts round the bow, flyingfish numerous, and Pyrosoma being used as shelter by little fishes.

Noon position: Lat. 2° 25′ N: Long. 86° 06′ W.

June 18th. Calm day, steaming steadily, except for one evening haul from 8:30 to 9:00 p. m., getting many *Myctophum coccoi* and two beautiful *Astronesthes.*

Noon position: Lat. 4° 27′ N: Long. 83° 44′ W.

June 19th. One evening surface haul again brought in *Myctophum coccoi,* and four *Astronesthes,* two of which had swallowed a large *coccoi* entire. This feeding habit accounts for their constant presence with the schools of *Myctophum.*

Noon position: Lat. 6° 10′ N: Long. 81° 33′ W.

June 20th. Caught a two-foot Coryphæna on spoon from boom-walk while ship was going at full speed.

Arrived Balboa late at night, to be met by customs launch down the harbor and boarded by new recruits.

Noon position: Lat. 7° 52′ N: Long. 79° 43′ W.

June 21st to 26th. Ship undergoing necessary overhauling for condenser and so on, and re-coaling. Official calls and dinners, and a day's excursion to Taboga Island, where we swam, walked, ate and tried the diving helmet but found the water so murky compared to the places we had been going down that it was not much use.

Sailed early on the morning of the 26th, but anchored in the Bay to wait for a delinquent fireman. Firemen always seem to exert a maleficent influence on our sailing days. At last got under way for Puerto Bello, entering the harbor and anchoring about 2:30 p. m.

Most interesting place, the ruins of the forts, that have been stormed so many times, in a surprisingly good state of preservation, with battlements, sentry-boxes and belfries, sloping ramps by which to drag the cannon to the walls, dungeons and barracks, and many cannon, all spiked.

The present town has about three hundred inhabitants, living in shaky houses that are mostly built on the fine solid foundations of the old buildings. The old barracks or customs house is quite a palatial structure, of several stories built in successive arches, roofless now and overgrown inside and out with natural window-boxes of bushes and flowers. San Blas Indians paddled alongside the *Arcturus* in dug-outs to inspect the ship at close range, and the people of the town proved to be a mélange of Panamanians, Indians, French half-breeds and a Finn who is a sergeant in the American army.

Aboard at 5 p. m. and out into the Caribbean.

Noon position: Lat. 9° 37′ N: Long. 79° 42′ W.

June 27th. Plowing through a grey sea, with a head wind, foul ship's bottom and everything combined to make it look as though we might spend the rest of the summer crossing the Caribbean. Now on our way to have another look for the Sargasso Sea, hoping to find a calmer Atlantic than we encountered in February.

Noon position: Lat. 10° 40′ N: Long. 78° 56′ W.

June 28th. Bright day, with the ship doing some fairly good nose-dives into big seas. From a mile off the Panama coast we have been passing weed in thousands of small pieces, little of it fresh, but mostly dark and discolored. The only signs of life aside from that sheltered in the weed are hundreds and thousands of flyingfish, almost all medium or small, and clear-winged.

Noon position: Lat. 12° 24′ N: Long. 77° 33′ W.

June 29th. Heavy head wind and seas, progress pitifully slow, making about three knots part of the day. From the boom-walk we can see the crusted growth of weed and barnacles below the water-line every time the ship rolls, and it is wonderful that we make any headway with such a handicap.

Noon position: Lat. 13° 41′ N: Long. 77° 06′ W.

June 30th. Same sort of weather. Ship plowing bravely on. We seem to be holding our own, anyway, but not much more. Evening reading-aloud-with-discussion-parties have become a regular feature.

Noon position: Lat. 14° 55′ N: Long. 76° 08′ W.

July 1st. Calmer to-day, making better time. Sighted Navassa Light this evening and passed it soon after midnight, in wonderful moonlight. First birds seen, several laughing gulls, a shearwater, and three yellow-billed tropicbirds.

Noon position: Lat. 16° 46′ N: Long. 74° 50′ W.

July 2nd. Passing between Cuba and Haiti. A perfect day, making our best speed, which is not a dizzy one. Small amount of weed seen, and worked from bow and pulpit collecting it. Three distinct species found. In strong contrast with our experience in February and March, almost every piece shelters fish and other animals. But all the weed showed signs of long submersion, much decay and extensive encrustations of dead animal life.

Passed Cape Maysi Light, eastern point of Cuba, about 7 p. m.

Noon position: Lat. 19° 24′ N: Long. 74° 20′ W.

July 3rd. For two nights there has been a tremendous display of sheet lightning, originating in three or four places and illuminating the whole sky and sea.

Perfect weather to-day, while we wound round islands and bays, among innumerable shoals and lagoons and lights. The traffic was heavy, four or five ships at a time in sight almost all day. Passed between Crooked Island and Long Island in the middle of the day. They are typical tropical islets of white sand beaches, scattered palms, low vegetation on the slopes of dunes, a banded red-and-white lighthouse and a lagoon of the clearest, greenest water imaginable.

At 8 p. m. made our first haul in a long while, getting quantities of weed and many species of small fish. Weed passing all day in very small pieces.

Noon position: Lat. 22° 09′ N: Long. 74° 22′ W.

July 4th. Misguided patriots firing guns at dawn. Another lovely day. Everyone taking turns in pulpit scooping up weed, of which there are many small bits. Several species of fish, and a vast quantity of larval ones, which so far we cannot identify. Every fish on our list so far, (except of course a few Myctophids taken at night) may be considered as being at least indirectly associated with sargassum weed.

Caught porcupine and triggerfish swimming alone as well as sheltered in the weed.

Noon position: Lat. 24° 03′ N: Long. 73° 01′ W.

July 5th.   Early in the morning the shaft of the circulating pump broke.   Shifted to sanitary pump and went ahead at slow speed on both engines.   Crept along all day, making about two knots.   Fortunately the Atlantic is wonderfully calm, with gorgeous sunsets and dawns.

A Coryphæna harpooned from pulpit to-day, stomach crammed with triggerfish.

Noon position: Lat. 25° 07′ N: Long. 71° 36′ W.

July 6th.   Half speed all day, with wind and sea against us, while repairs are made on pump.   Weed passing steadily in small, scattered patches.

Noon position: Lat. 25° 56′ N: Long. 70° 31′ W.

July 7th.   Half speed or stopped completely all day.   An increase in weed to-day, especially in size of pieces, which were rounded heads of new, pale, freshly grown sargassum.   Trying a new sort of net, which caught much weed and many small triggerfish, *Xenichthys ringens.*

About 6 p. m. pump-shaft repaired and we started full speed for 30° North and 60° West, hoping to find abundance of weed and good weather.

Noon position: Lat. 25° 54′ N: Long. 69° 30′ W.

July 8th.   Gregory and Beebe saw a remarkable-looking shark from the deck this morning.   It was about six feet long and seemed to be much more heavily built than the common sharks.   It swam so close to the surface that its dorsal fin protruded and its color seemed to be a pale sage green, with at least half the dorsal and pectoral clear white.   It swam slowly past, then turned and followed the ship for a time.

Noon position: Lat. 26° 45′ N: Long. 67° 56′ W.

July 9th.   We have lost the trades to-day.   Weed more abundant and considerable new growth in compact heads.   Saw Physalia, several *Pterophryne,* pipefish and triggerfish.   Evening tows with half-meter net brought in only eight *Myctophum coccoi* and much weed.

Noon position: Lat. 27° 51′ N: Long. 65° 26′ W.

July 10th.   Perfect day,—the sea a mirror before sunrise and only lightly ruffled later.   Less weed than ever.   Two tropicbirds in early morning, perhaps our friends who were so persistent in February!   Evening tow yielded little.

Noon position: Lat. 28° 54′ N: Long. 62° 50′ W.

July 11th. Arrived at 30° North and 60° West, to make Station 96.

The weed for to-day was far below the average of last three days. One unusual condition was that a large percentage of it was submerged quite deeply, appearing in good sized patches and new, but so obscured by the water that nothing positive could be asserted of the pieces.

Reached our chosen spot at 2 p. m. and sounded in 2875 fathoms. This is Saturday, but we have declared it to be the crew's Sunday, so that they can put in a full day's work to-morrow.

A fairly calm sea, with practically no weed in sight, and several small rain-storms passing close by. Everyone went swimming.

Noon position: Lat. 29° 53′ N: Long. 60° 18′ W.

July 12th. This seems something like a doldrum at last,—a smooth blue sea, and scarcely a breath of breeze, and what there is, is hot. Put down a Petersen and five meter nets and got less than in one net in the Pacific. The best specimens were a small dark brown *Saccopharynx* and a two-inch *Remora*, with many larval characters.

Noon position: Lat. 30° 01′ N: Long. 60° 03′ W.

July 13th. Stopped almost all day for vertical hauls.

Noon position: Lat. 30° 42′ N: Long. 61° 16′ W.

July 14th. Another day of vertical hauls.

A block came down on John's head, slicing his scalp in three places, and Jay ran a rusty spike in his foot, and is now on crutches after a tetanus injection.

Noon position: Lat. 31° 22′ N: Long. 62° 35′ W.

July 15th. Hard rain last night. Chiriqui's cage blew off the library deck and only by a miracle did not go completely overboard.

Sighted Bermuda this afternoon amid squalls and skirted it during dinner. Contrast with approach to one of our Pacific islands is strong, —no boobies coming out to meet us, no sea-lions bobbing round us, and the only life a few red-billed tropicbirds.

Towed this evening and got only four small *Myctophum coccoi*. All our observations in the Sargasso Sea show small individuals and few species compared to the Pacific fauna.

Reached the vicinity of Challenger Bank in evening.

Noon position: Lat. 32° 06′ N: Long. 64° 17′ W.

July 16th. Sounded in 25 fathoms, thirteen miles from land, which is low on the horizon. Put out a big alcohol drum painted white to serve as a guide and anchored it with 60 fathoms of rope.

Made two fifteen-minute dredges on the bank and both times had the nets badly torn. Almost no life but an abundance of large grey gorgonias, and many kinds of algæ, such as thick heads, masses of small round egg-like forms, a sargassum with flattened berries, flat lettuce stuff, pale grey green with purple edges, plume shapes, masses of forms like threads, etc.

Several large hard head sponges, and some slender, hollow, finger-shaped ones. In the big sponges were brittle stars only, deserving their name as they fell to pieces at a touch. There was one blood-red large one and several small scarlet squilla.

In both dredges we got only two small fish, one of which was a young moray eel.

When we chiselled apart the great solid coral boulders, life was found to be more abundant and we captured numbers of creatures, including ascidians, etc.

We next went to the eastward of the buoy and a mile away still found bottom at 28 fathoms; another mile and we got 505 fathoms, and put down a Petersen and three metre nets in 300, 400, and 500 fathoms. In two and a half hours we got a good haul.

Echiostoma barbatum was over a foot long, soft and flabby but not with the tender skin of the deep-sea forms. It was alive and stayed so for several hours while we got movies. The most noticeable character of this otherwise brownish-black fish was a wedge- or pear-shaped light organ of rich rose color below the eye. In the dark this gave forth a warm reddish glow. The lateral light organs were all tinged with rose.

One of the strangest creatures of the entire trip was a small jetfish, *Caulophryne jordani,* a little rounded brownish black creature, with outrageously long dorsal and anal fin rays, and covered from lips to tail with fleshy tactile filaments. He was at his last gasp, unfortunately, but we got sketches and photographs.

In the evening from 8 to 9 p. m. towed two metre nets at 200 and 300 fathoms, obtaining only a meagre haul but showing an interesting elevation of red forms nearer the surface than in the daytime. This evening also we captured in one of these nets the deep-sea prawn that throws off a cloud of luminescence as a cloak behind which it escapes its enemies.

Noon position: Lat. 32° 02′ N: Long. 65° W.

July 17th. Put down eight nets from 100 to 800 fathoms, and got a very satisfactory plankton haul, but the fish were almost wholly young or immature forms. Sea which kicked up in the afternoon made pulling in a difficult matter. Started west.

Noon position: Lat. 31° 57′ N: Long. 64° 55′ W.

July 19th. Heavy seas and a staggering wind, with an occasional wave coming over the port bow in a smother of spray. Difficulty in taking temperatures this morning.

Noon position: Lat. 33° 14′ N: Long. 68° 35′ W.

July 20th. Opposite of yesterday,—smooth blue sea and baking sun. Practically no weed, only very small pieces here and there. Several hours in pulpit brought in only two pailsful, containing two young *Pterophryne.* Temperature at 6 a. m. showed no signs of Gulf Stream, though we must be near the edge. At 1 p. m. put over five metre nets at 250, 500, 600, 700 and 800 fathoms, and got a fairly good haul (for the Atlantic). As usual most of the Cyclothones and *Argyropelecus* were very young. One splendid large black fish with scores of pink body lights, and a very broad mid-side band of bronze and copper extending down body from head to tail.

In the afternoon hooked a six-foot shark, *Carcharias milberti,* with one shark sucker attached (*Remora remora*) and many big parasitic Copepods.

Noon position: Lat. 34° 19′ N: Long. 71° 13′ W.

July 21st. Another wonderful day. A large clear-winged dragonfly seen this morning. Five metre nets down to 250, 500, 600, 700 and 800 fathoms (Station 107) at 1 p. m.

Wireless from Harrison Williams that he is on board the *Warrior* off Hampton Roads and will meet us to-morrow. At 5 p. m. we steamed north, making 10 knots with the wind and Gulf Stream to help us.

Noon position: Lat. 34° 47′ N: Long. 73° 41′ W.

July 22nd. At 6 a. m. a densely massed flock of over three hundred stormy petrels feeding close by, to starboard. They kept just ahead for a hundred yards, and then broke up and drifted astern, only a few keeping on with us.

Rather rough sea with whitecaps. At 7 a. m. we effected a perfect meeting with the Warrior, and her passengers, five in number, came aboard for lunch and to watch us make a haul with five metre nets. Fair results. In the afternoon two of the visitors returned to the yacht, but it was so rough that the others stayed on the *Arcturus,* and we followed the *Warrior* toward Chesapeake Bay. Big wind and rain to-night, foghorn bellowing.

While passengers were being transferred, a United Fruit Liner on the horizon altered her course and passed within a few hundred yards of us. Two ships tossing aimlessly about, while lifeboats labored in the waves, made a picture that evidently required investigation.

Noon position: Lat. 36° 53′ N: Long. 74° 10′ W.

July 23rd. Early this morning the *Warrior* returned to our side. We put down two dredges and obtained one starfish, one clam and a banana peel! Not a rich territory.

All went aboard the *Warrior* for lunch, and at 3 p. m. parted, the *Arcturus* headed for Lat. 38°, Long. 74°.

Noon position: Lat. 36° 56′ N: Long. 75° 25′ W.

July 24th. Arrived at our station about 7 a. m. and lowered a dredge which evidently did not touch bottom, as it came up with nothing but several bushels of salpæ. The second dredge was lowered to 500 fathoms, and while being towed, struck something that stopped the ship as though she had run aground. When recovered, after much manœuvring, the dredge was torn almost off the frame, and was bulging with what looked like a ton of green mud, into which every-one plunged to sort out anemones, starfishes, holothurians, worms, crabs, beautiful purple pennatulas. The fish were mostly hakes, with one large goosefish, some Scorpænids and a slender *Lycenchelys*.

The dredge stuck because we were pulling against the continental slope, beginning at 500 and ending at about 250 fathoms, so that the net and frame were buried in bottom mud.

Now going to the Hudson River Gorge, submerged since the Pleistocene, where we plan to do some dredging.

Noon position: Lat. 38° 11′ N: Long. 73° 44′ W.

July 25th. Dredging started directly after breakfast. The big dredge brought up nothing at all. Then put out five metre nets and got a good haul, with Myctophids, Cyclothones, *Stomias, Chauliodus, Serrivomer, Nemichthys, Caulophryne,* a blue-eyed flounder, etc.

The dynamite squad went over in a small boat and used up the last charge, under a bit of wreckage, but every fish sank so rapidly after the explosion that they could not be recovered.

At 10 p. m. began a twenty-four-hour haul of half-metre surface nets. A few Myctophids, but most of the fish are young hakes.

During evening tried for fish by lighted gangway. Got squid, a snake mackerel, and beautifully marked Nomeus.

Noon position: Lat. 39° 13′ N: Long. 72° 12′ W.

July 26th. Steamed slowly all day, making the twenty-four-hour plankton hauls. A cold grey windy day.

Noon position: Lat. 39° 26′ N; Long. 71° 26′ W.

July 27th. Vertical and closing-net hauls all the morning. Crew catching sharks; got three in about half an hour, and we had a steak from one for lunch. Very good, not unlike swordfish but tougher and not so dry. Crew vastly disgusted at the idea, but cannot explain why.

Nets in the afternoon, with fair results.

Noon position: Lat. 39° 37′ N: Long. 71° 40′ W.

July 28th. Calm and fine. Made a deep haul this morning and another this afternoon. Every net bulging with quantities of salpæ, but in the mass of jelly, when floated out in tubs and dishes, were some interesting fish,—*Caulophryne, Serrivomer, Melamphæs, Oneirodes* and many very large Cyclothones.

Many dolphins leaping in the middle distance, and we have seen blackfish and tunnies. This morning a spearfish, six or eight feet long, swam close under the boom, where the trolling line was trailing, and once even took the spoon in its mouth, but was not hooked.

Noon position: Lat. 39° 16′ N: Long. 71° 40′ W.

July 29th. A day of dismantling, packing, and concentrated industry.

Noon position: Lat. 39° 43′ N: Long. 72° 15′ W.

July 30th. Arrived in the lower bay at dawn, and proceeded to berth at 81st Street and North River (Fig. 68). Every ship on the way saluted us, from garbage scows to big liners, and as courtesy required that we answer every blast, we had barely steam enough to creep up to the pier.

APPENDIX A

NOTES ON THE FAUNA OF OSBORN ISLAND

BY WILLIAM BEEBE

Osborn Island, in honor of Prof. Henry Fairfield Osborn, is the name I have given to the larger of two islands lying between Hood and Gardner. While I plan later to publish the details of my study of the fauna of this interesting island, I here offer an annotated list of what I believe is its complete avifauna at the time of my visit, with notes on other groups. I spent four hours on the island in the forenoon of April 27th, 1925. The configuration of the island is such that a complete census was possible in this very brief time. Of the birds, there were 17 species, represented by 67 individuals.

BIRDS OF OSBORN ISLAND

1 *Diomedea irrorata* Salvin
Galápagos Albatross
Five of these flew so close to the island that they passed over the rocky promontory where the sea-lions were basking.

2 *Sula nebouxi* Milne-Edwards
Blue-footed Booby
Two pairs of these birds roosting, and perhaps breeding on the ledges of an inaccessible part of the cliff.

3 *Sula piscatrix websteri* Rothsch.
Red-footed Booby
Three birds flew over and one alighted for a moment near the Blue-foots.

4 *Pelecanus fuscus californicus* Ridgway
  Galápagos Brown Pelican
  One flying low over island.

5 *Fregata* sp.
  Frigatebird
  Two soaring high over the island.

6 *Creagrus furcatus* (Neboux)
  Fork-tailed Gull
  Four pairs nesting on cliffs on west side of island.
  Several nestlings seen.

7 *Blasipus fuliginosus* (Gould)
  Lava Gull
  Three walking about on the beach.

8 *Anous stolidus* (Linné)
  Noddy Tern
  Several pairs nesting on ledges and others flying about.

9 *Butorides sundevalli* Reichenow
  Galápagos Green Heron
  A single individual catching flies near the shore.

10 *Nesopelia g. galapagoensis* (Gould)
   Galápagos Dove
   Three pairs on the island, one of which soon flew to
   Gardner. Collected one of the others in a butterfly
   net as it walked up within a yard of me. It was a
   breeding female. One pair kept flying to a considerable height above the island and then soaring slowly to
   the ground,—quite an undovelike manœuvre.

11 *Buteo galapagensis* (Gould)
   Galápagos Hawk
   One bird soaring overhead.

12 *Eribates magnirostris* (Gray)
   Yellow-bellied Flycatcher
   Two followed me about the island, all but alighting
   on my shoulder, and showing great interest whenever I
   turned over a stone in my search for geckos.

13 *Certhidea cinerascens* Ridgway
    Gray Certhidea

Two pairs nesting on the island, one on each side of the central ridge. The first nest was complete, two feet from the ground, large and globular, and contained a single egg. I shot the female and collected the nest, as the egg of this species is still undescribed. The second pair was still building, the nest almost completed and composed wholly of moss. Both males were singing, and although quite out of sight of one another, the songs alternated, the one on the east of the ridge waiting until the song of the bird on the other side was ended. The song was of the simplest, a sibilent *Sip-sip-sip-sip—chew—chew—chew!*

14 *Dendroica petechia aureola* (Gould)
    Galápagos Yellow Warbler

One of these warblers hopped about the bushes near me for a while and then flew away in the direction of Gardner.

15 *Geospiza conirostris* Ridgway
    Conebilled Ground Finch

Two pairs on the island, one nesting. There were two nests, one old, evidently last year's, and one just finished. Both males were singing from the top of a bush or cactus, and taking short flights between songs, often followed by the females. The notes were of the simplest, but given with great impetus, in a hoarse voice, *Chuckel-low!* The eggs of this species are also unknown, but the nest was empty.

16 *Geospiza fuliginosa* Gould
    Sooty Ground Finch

There were five pairs on the island, two building nests, one with a full-grown young, stuffing it with small, green measuring worms, which had almost defoliated the bursera trees. I shot the male of a pair wandering through the underbrush. Its mate saw it fall, looked at it for a moment, and then went on searching for food. The song bore a remarkable resemblance to that of the Certhidea. It may be written, *Sip-sip-sip-sip—seep—seep—seep!* the first four syllables uttered very rapidly.

17 *Nesomimus macdonaldi* Ridgway
   Hood Island Mockingbird
   One individual, perched on a shrub, flew off before I could approach closely.

OTHER VERTEBRATES OF OSBORN ISLAND

Phyllodactylus bauri Garman
   Hood Island Gecko
   Collected four and saw three others. All were under stones, and with each was a little pile of the elytra of the beetle *Stomion.*

Tropidurus delanonis Baur
   Hood Island Lizard
   At least eight of these large lizards seen running over the ground. All had an unusual amount of red on the body, and a small one which I collected was almost wholly of this color.

Amblyrhynchus cristatus Bell
   Marine Iguana
   Saw three medium-sized specimens, all with a great deal of brick color on the scales.

Otaria jubata (Gmelin)
Southern Sea-lion
   About a dozen in the cave, and sixteen basking at one time on the rocky, southeast peninsula. All of this latter group were affected with conjunctivitis, some of the pups being quite blind.

APPENDIX B

TEXT IDENTIFICATIONS

APPENDIX B 435

436   APPENDIX B

PAGE	LINE	
309	9	White-striped Angelfish — *Holocanthus passer* Valen.
310	3	*Hepatus triostegus* (Linné)
310	10	*Pachygnathus capistratus* (Shaw)
311	6	Moorish Idol—*Xanclus cornutus* (Linné)
311	14	Parrotfish—*Scarus* sp.
324	31	Fairy Tern—*Gygis alba candida* (Gmelin)
328	5	Red-footed Booby — *Sula piscatrix websteri* Rothsch.
328	5	White-breasted Booby—*Sula brewsteri* Goss
330	10	Lava Gull—*Blasipus fuliginosus* (Gould)
330	16	Noddy Tern—*Anous stolidus ridgwayi* Anthony
330	16	Cocos Noddy Tern—*Megalopterus minutus diamesus* H and S
330	28	Yellow Warbler — *Dendroica petechia aureola* (Gould)
331	23	Sea Turtles—*Chelone mydas* (Linné)
332	1	Pacific Dolphins — *Lagenorhynchus obliquidens* Gray
332	25	Cocos **Brassolid** Butterfly—*Historis orion* (Fabricius)
354	21	Flying Snails—Pteropods such as *Clio* and *Cavolinia*
354	28	Black Octopus—*Cirroteuthis* sp.
359	7	Dwarf Sharks—*Centroscyllium nigrum* Garman
359	24	Rainbow Snails—*Gaza rathbuni* Dall
345	28	Scarlet Luminiscent Shrimp—*Pandalus annulicornis* Leach
360	11	Pelicanfish—*Saccopharynx* sp.
360	12	Thread Eel—*Nemichthys* sp.
361	13	Anglerfish—*Lophius piscatorius* Linné
376	15	Red-tailed Triggerfish — *Xanthichthys ringens* (Linné)
377	21	Stormy Petrels—*Oceanites oceanicus* (Kuhl)
379	10	Snake Mackerel—*Gempylus* sp.
382	23	Scorpionfish—*Helicolenus maderensis* G and B

INDEX

437